Memorandum

Dear Reesor,

Alan O'Connor mentioned you might like to review our book. It is a bit on the boring side, but does a really good job of seeing the foundation for how to design programs that align and

Energy-Based Economic Development

Sanya Carley · Sara Lawrence

Energy-Based Economic Development

How Clean Energy can Drive Development and Stimulate Economic Growth

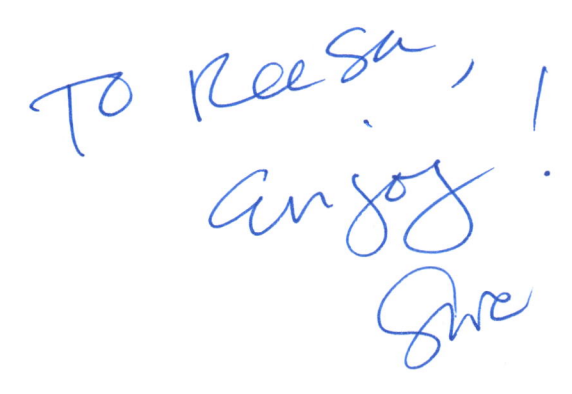

Springer

Sanya Carley
Bloomington, IN
USA

Sara Lawrence
Raleigh, NC
USA

ISBN 978-1-4471-6340-4   ISBN 978-1-4471-6341-1   (eBook)
DOI 10.1007/978-1-4471-6341-1
Springer London Heidelberg New York Dordrecht

Library of Congress Control Number: 2014932992

© Springer-Verlag London 2014
This work is subject to copyright. All rights are reserved by the Publisher, whether the whole or part of the material is concerned, specifically the rights of translation, reprinting, reuse of illustrations, recitation, broadcasting, reproduction on microfilms or in any other physical way, and transmission or information storage and retrieval, electronic adaptation, computer software, or by similar or dissimilar methodology now known or hereafter developed. Exempted from this legal reservation are brief excerpts in connection with reviews or scholarly analysis or material supplied specifically for the purpose of being entered and executed on a computer system, for exclusive use by the purchaser of the work. Duplication of this publication or parts thereof is permitted only under the provisions of the Copyright Law of the Publisher's location, in its current version, and permission for use must always be obtained from Springer. Permissions for use may be obtained through RightsLink at the Copyright Clearance Center. Violations are liable to prosecution under the respective Copyright Law.
The use of general descriptive names, registered names, trademarks, service marks, etc. in this publication does not imply, even in the absence of a specific statement, that such names are exempt from the relevant protective laws and regulations and therefore free for general use.
While the advice and information in this book are believed to be true and accurate at the date of publication, neither the authors nor the editors nor the publisher can accept any legal responsibility for any errors or omissions that may be made. The publisher makes no warranty, express or implied, with respect to the material contained herein.

Printed on acid-free paper

Springer is part of Springer Science+Business Media (www.springer.com)

Acknowledgments

We gratefully acknowledge Brian Southwell, Vikram Rao, Patricia Scruggs, Daniel Raimi, Mark Skinner, and Ted Abernathy for peer-reviewing drafts of the book. We also thank the three anonymous reviewers of the original proposal for their constructive feedback.

Several research assistants provided valuable support. We thank Marty Hyman, Eric Fisher, Laura Nicholson, Elinor Benami, Andrew Nourafshan, Michael Spolum, Tingting Tang, Rachel Dimmit, and Ope Onibokun.

We are very appreciative of the thoughtful contributions from Mr. Myles Elledge on the national case studies selected in Chap. 8 and Mr. Daniel Raimi for his research on natural gas development in Southwestern Pennsylvania. Mr. Raimi's master's thesis *The Potential Social Impacts of Shale Gas Development in North Carolina* offered helpful insights and knowledge about natural gas development for that sub-national case study.

We especially thank Dave Myers, Vikram Rao, and Alan O'Connor for encouraging us to write this book. Thank you also to Sharon Barrell, Lee Anne Nance, Pernille Dagø, Sara Casey, and David Chrest for helping us collect information or complete different aspects of the book.

This book benefited considerably from the contributions of coauthors that helped us write previous energy-based economic development publications, including Adrienne Brown, Sameeksha Desai, Morgan Bazilian, Daniel Kammen, Elinor Benami, Marty Hyman, and Andrew Nourafshan.

Finally, we would like to thank our spouses, Joe and Rick, for their unwavering support.

Contents

1	**Energy-Based Economic Development**		1
	1.1 Why EBED, Why Now?		2
	1.2 Reaching a Diverse Audience		4
	1.3 Complex Challenges that Establish the Need for EBED		5
		1.3.1 Reduce Greenhouse Gases and Other Emissions	5
		1.3.2 Improve Cost, Efficiency, and Energy Security	6
		1.3.3 Reduce Energy Poverty	8
		1.3.4 Identify Energy-Based Economic and Employment Opportunities	9
		1.3.5 Identify Energy Links to Alleviate Poverty	9
	1.4 Outline of the Book		10
	References		12
2	**Defining Energy-Based Economic Development**		15
	2.1 Definition		15
		2.1.1 Other Definitions of the Energy–Development Nexus	17
	2.2 Foundations		19
		2.2.1 Economic Development	19
		2.2.2 Energy Policy and Planning	20
		2.2.3 Convergence	22
	2.3 EBED Goals		23
		2.3.1 Energy Goals	23
		2.3.2 Economic Development Goals	27
	2.4 Conclusion		32
	References		32
3	**Process and Approaches**		37
	3.1 Process		38
		3.1.1 Engage Stakeholders	40
		3.1.2 Identify Goals and Objectives	40
		3.1.3 Identifying Assets, Needs, and Gaps	41
		3.1.4 Select and Design Strategy and Approach	43
		3.1.5 Identify Metrics	44

		3.1.6	Pilot and Implement	44
		3.1.7	Monitor and Evaluate	45
	3.2	EBED Approaches		45
		3.2.1	Point of Intervention	45
		3.2.2	Geographic Scale	49
		3.2.3	Scale of Transformation	51
	3.3	Conclusion		52
	References			53
4	**Supportive Policies for Energy-Based Economic Development**			**55**
	4.1	Technology Innovation Policies		58
	4.2	Technology Adoption and Commercialization Policies		60
		4.2.1	Feed-In Tariffs	60
		4.2.2	Net Metering, Interconnection Standards, and Framework Laws	61
		4.2.3	Loan Guarantees	62
		4.2.4	Incentives	62
		4.2.5	Government Procurement and Demonstration	63
		4.2.6	Information and Education	64
		4.2.7	Regulatory Standards	64
	4.3	Entrepreneurship Policies		65
		4.3.1	Start-Up and Expansion Capital	66
		4.3.2	Access to Infrastructure and Services	67
		4.3.3	Entrepreneurship Awareness and Training	67
	4.4	Industrial Growth Policies		68
		4.4.1	Business Climate Policies	70
		4.4.2	Information and Coordination Policies	70
		4.4.3	Import Substitution, Export Promotion, and Foreign Direct Investment Policies	71
		4.4.4	R&D for Industrial Growth Policies	72
	4.5	Workforce Development		73
	4.6	Climate and Environmental Policies		74
		4.6.1	Emission Performance Standards	74
		4.6.2	Direct Emissions Regulation	74
		4.6.3	Taxes and Cap-and-Trade Programs	75
	4.7	Planning		75
		4.7.1	Integrated Resource Planning	76
		4.7.2	Comprehensive and Strategic Planning	76
		4.7.3	Low Emission Development Planning	77
		4.7.4	Sustainable Cities Planning	77
	4.8	Conclusion		78
	References			78

5	**Evaluation and Metrics**.		81
	5.1	Outcome Metrics	82
	5.2	Type of Initiative Evaluated.	86
	5.3	Methodological Approach	87
	5.4	Timing and Research Design	89
	References		90
6	**Case Study Approach**.		95
	6.1	Selection of EBED Cases	96
7	**Subnational EBED Cases**.		99
	7.1	Case Study 1: The Bandeirantes Landfill Gas to Energy Project	100
		7.1.1 The Program	100
		7.1.2 EBED Framework	101
	7.2	Case Study 2: Clean Energy Works Oregon.	102
		7.2.1 The Program	103
		7.2.2 EBED Framework	103
	7.3	Case Study 3: Copenhagen Cleantech Cluster	104
		7.3.1 The Program	105
		7.3.2 EBED Framework	106
	7.4	Case Study 4: Kamworks, Rural Cambodia	106
		7.4.1 The Program	107
		7.4.2 EBED Framework	108
	7.5	Case Study 5: Natural Gas Development in Southwestern Pennsylvania, United States	109
		7.5.1 The Program	110
		7.5.2 EBED Framework	112
	7.6	Case Study 6: Nuru Energy	113
		7.6.1 The Program	113
		7.6.2 EBED Framework	114
	References		115
8	**National Case Studies**.		119
	8.1	Case Study 7: Biofuels in Singapore.	119
		8.1.1 The Program	120
		8.1.2 EBED Framework	121
	8.2	Case Study 8: China Golden Sun	122
		8.2.1 The Program	122
		8.2.2 EBED Framework	123
	8.3	Case Study 9: Ethiopia National Cookstoves Program.	124
		8.3.1 The Program	125
		8.3.2 EBED Framework	125

	8.4	Case Study 10: Lao People's Democratic Republic National Hydropower Initiative	126
		8.4.1 The Program	127
		8.4.2 EBED Framework	127
	8.5	Case Study 11: Morocco Solar and Wind	128
		8.5.1 The Program	129
		8.5.2 EBED Framework	130
	8.6	Case Study 12: South African Renewables Initiative	131
		8.6.1 The Program	132
		8.6.2 EBED Framework	133
	References		134
9	**A Hybrid Model: The American Recovery and Reinvestment Act**		**137**
	9.1	Overview of ARRA	138
	9.2	Energy-Related Recovery Act Offices and Programs	139
		9.2.1 Office of Energy Efficiency and Renewable Energy	141
		9.2.2 Office of Electricity Delivery and Energy Reliability	145
		9.2.3 DOE Loan Programs Office	145
		9.2.4 Department of Housing and Urban Development's Green Retrofit Program	146
		9.2.5 Commonalities Within Programs	147
	9.3	Case Studies of Selected Local Recovery Act–Funded Initiatives	147
		9.3.1 Green Launching Pad	147
		9.3.2 Energize Phoenix	148
		9.3.3 Summary of Case Studies	150
	9.4	Early Evaluations of ARRA and Potential Implications	150
	9.5	Conclusions	152
	References		153
10	**Common Themes and Conclusions**		**157**
	10.1	EBED Efforts Often Require a Multidimensional and Comprehensive Approach	158
	10.2	There is no Single Prescription	158
	10.3	Timing is Crucial and Difficult	159
	10.4	Strategic Investment may be Necessary	160
	10.5	Project Self-Sufficiency can be Challenging	161
	10.6	Public–Private Partnerships Play an Important Role	161
	10.7	Attention to Economic Benefit and Burden is Important	162
	10.8	EBED Efforts may be Met by Unintended Consequences	162

10.9	Political Will and a Consistent, Stable Policy Environment is Crucial	163
10.10	Community Participation is Important, Especially for Place-Based Approaches	163
10.11	EBED in the 21st Century	164

Acronyms

AGECC	Advisory Group on Energy and Climate Change
ARPA-E	Advanced Research Projects Agency-Energy
BLS	Bureau of Labor Statistics
CCS	Carbon Capture and Sequestration
CNG	Compressed Natural Gas
CO_2	Carbon dioxide
CSIR	Council on Scientific and Industrial Research
CUSP	Curtin University Sustainability Policy Institute
EBED	Energy-Based Economic Development
EE	Energy Efficiency
EIA	Energy Information Administration
EUNIP	European Network on Industrial Policy
FDI	Foreign Direct Investment
GDP	Gross Domestic Product
GHG	Greenhouse Gas
GW	Gigawatt
HOV	High Occupancy Vehicle
IBM	International Business Machines
IEA	International Energy Agency
ILO	International Labour Organization
IOE	International Organisation of Employers
IPCC	Intergovernmental Panel on Climate Change
ITUC	International Trade Union Confederation
JEDI model	Jobs and Economic Development Impact model
JSBC	Japan Small Business Corporation
kW	Kilowatt
kWh	Kilowatt hour
LED	Light-emitting Diode
LEDS	Low Emission Development Strategies
LFG	Landfill-derived Gas
m^2	Square meters
M-PESA	Mobile money (Swahili)

MW	Mega-watt
MWa	"Average installed megawatts de-rated by the capacity factor of the technology" (Kammen et al. 2006)[1]
MWh	Megawatt hour
NASA (in refs)	National Aeronautics and Space Administration
NGA	National Governors Association
NGO	Nongovernmental Organization
NJPIRG	New Jersey Public Interest Research Group
NOx (or NO$_x$)	Nitrogen oxides
NREL	National Renewable Energy Laboratory
O&M	Operations and Maintenance
OECD	Organisation for Economic Cooperation and Development
Open EI	Open Energy Info
R&D	Research & Development
RD&D	Research, Development, and Deployment
RE	Renewable Energy
REPP	Renewable Energy Policy Project
RES	Renewable Energy Standard
RTI	RTI International
SARi	South African Renewables Initiative
SBIR	Small Business Innovation Research
SMEs	Small to Medium Enterprises
SO2 (or SO$_2$)	Sulfur dioxide
TIP	Technology Information Policy Consulting
TVEs	Township and Village Enterprises
TWh	Terawatt hour
US	United States
(US) EPA	Environmental Protection Agency
UN	United Nations
UNDP	United Nations Development Programme
UNEP	United Nations Environmental Programme
UN-Habitat	United Nations Human Settlements Programme
USA	United States of America
WB	World Bank
WHO	World Health Organization

---

[1] Kammen D, Kapadia K, Fripp M (2006) Putting renewables to work: how many jobs can the clean energy industry create? Renewable and Appropriate Energy Laboratory (RAEL), University of California, Berkley, CA.

Chapter 1
Energy-Based Economic Development

Abstract In recent years, energy has become much more of a driver of new paths to economic development than an enabler of growth. Accordingly, governments and the private sector are investing billions of dollars annually in low-emissions energy development and energy efficiency planning. From 2004 to 2010, global renewable energy development increased 540 %. This chapter introduces the domain of energy-based economic development (EBED) and explains how it seizes joint opportunities inherent in energy development and economic development. This chapter also reviews the complex and overlapping issues that EBED most commonly addresses, including greenhouse gas and other emissions, energy security and efficiency, energy poverty, economic growth and recovery, job creation, and poverty.

Energy-based economic development (EBED) refers to efforts that simultaneously pursue energy policy and planning goals and economic development and growth goals. EBED is a growing discipline that takes advantage of the economic development opportunities inherent in low-emissions energy and energy-efficient development to generate new, innovative economic growth.

EBED activities are increasingly prevalent around the world with billions of dollars being invested by the public and private sectors. In 2010, governments and the private sector globally invested $211 billion in renewable energy development, a 32 % increase from 2009 and a 540 % increase from 2004. More than half of this investment was in large-scale renewable energy projects in developing countries (Frankfurt School—UNEP Collaborating Centre for Climate & Sustainable Energy Finance 2011). Various stimulus programs injected approximately $463 billion into energy and environmental projects between 2008 and 2012. Moreover, international development assistance for energy accounted for approximately $171.8 billion in 2011, with an average annual increase in funding of 10.2 % from 2000 to 2011 (Organisation for Economic Co-operation and Development [OECD] 2013). In addition, registered and operating energy-related projects under the

Clean Development Mechanism of the Kyoto Protocol received $89.2 billion in total as of 2012 (Kirkman et al. 2012).

The United States, China, and South Africa offer three examples of countries that have recently embraced EBED approaches to growth and development strategies. In early 2009, the American Recovery and Reinvestment Act (ARRA) devoted approximately $58 billion out of $840 billion to energy-related economic development projects. This massive economic stimulus effort, rivaled perhaps only by the New Deal, targeted energy efficiency, green jobs, smart grids, renewable energy, and advanced fossil energy, among a variety of other nonenergy related programs as well.

In 2009, China launched the "Ten Cities, Thousand Vehicles Program," in which the cities of Beijing, Shenzhen, Shanghai, Jinan, Chongqing, Wuhan, Changchun, Hefei, Dalian, and Hangzhou were encouraged to launch a test program to put over 1,000 electric vehicles on the roads. This program first targeted taxis, garbage trucks, and buses. The program soon expanded to other cities and to passenger vehicles. At the 2010 Beijing auto show, over 100 automakers showcased electric vehicles, either new or concept-stage models, reflecting the campaign by Chinese central planners to make China the first country to mass produce electric vehicles. The government plans to invest over $15 billion in 2013 and in subsequent years in research and development, subsidies, and recharging infrastructure and to establish several Chinese vehicle manufacturers as well as suppliers of specific electric vehicle components.

In November 2011, the South African Renewables Initiative (SARi) was formally established at the United Nations Framework Convention on Climate Change. SARi is a South African government-led initiative, managed by the Department of Trade & Industry and the Department of Energy. The primary objective of the initiative is to stimulate energy-based industrial activities through export competitiveness, renewable energy development, energy security, and job growth. One of the main components of this initiative is a plan to increase renewable capacity by approximately 18 GW per year and to eventually reach 15 % renewable energy between 2020 and 2025.

Together, these cases demonstrate an emerging global trend of an increasing focus on EBED. These trends reveal that it is an opportune time to evaluate how EBED programs work, common approaches employed, challenges implementers and administrators may encounter, and ways to evaluate these types of initiatives.

1.1 Why EBED, Why Now?

EBED is timely for several reasons. First, energy and economic development have always been linked, but in recent years energy has become much more of a *driver* of new paths to economic development rather than an *enabler* of growth. This distinction is subtle but important. Energy innovations are increasingly used as the primary vehicle for development, not just a factor of production.

Second, EBED involves advanced, low-emissions, and efficient energy, which is much more distributed in nature than more carbon-intensive energy resources such as coal. Renewable energy, natural gas production, and energy-efficient techniques are relevant across geographies and are scalable in ways that more traditional energy provision has not been in the past. The more dispersed nature of lower-emissions energy sources allows for a more diverse ownership of energy development. Thus, the distributive nature of energy is changing where energy can be developed and who benefits from these activities. Among other shifts, this phenomenon is not only making energy policy and planning relevant to a much boarder populace, it is also causing a change in the kinds of programs and policies currently deployed to support efforts in the energy field.

Third, EBED demands a multidisciplinary approach to tackling some of the more complex problems a community or country faces. Globally complex problems tend to involve a variety of actors, both those responsible for the problem and those involved in potential solutions, and differences in opinion about the most appropriate way to address the issue (Rittel and Webber 1973). Modern examples of complex problems include global climate change, food insecurity, water access, energy access, and poverty.

The interdependencies between these problems create an added degree of complexity that can proliferate into deeply intractable conditions. For example, droughts and other weather abnormalities caused by climate change affect subsistence crops in many regions. The inability of households that farm to feed themselves and their families has the potential to exacerbate food insecurity and to affect poverty and human health. Droughts also affect water supply and may render a hydroelectric dam that provides the bulk of a region's power useless and, thus, contribute to energy insecurity. Droughts can also limit access to clean water and affect food availability and health care provision. These kinds of complex problems are multilayered, having economic, environmental, social, and political dimensions, and require multidisciplinary rather than single-pronged approaches.

EBED offers one of many cross-disciplinary problem-solving approaches to these global challenges. In fact, EBED initiatives take an *opportunistic* approach to these challenges to exploit the potential for energy and development activities that produce positive economic, energy, social, and environmental outcomes.

Finally, with the proliferation of funding and emphasis on EBED there is a need to develop a common understanding among a diverse EBED audience. By definition, EBED is a process by which multiple stakeholders in a country or region strive to increase access to modern energy services, increase energy efficiency, improve energy governance, and diversify energy resources in ways that simultaneously generate industry growth, economic development, and national security.

Because the actors involved in EBED activities often come from different disciplinary backgrounds, EBED partnerships may not evolve organically. A common understanding and language about EBED, therefore, is necessary to create a unified framework for activities that occur in this field. We define the EBED domain to help forge this common understanding for the audiences involved—including practitioners, policymakers, and researchers—and to define shared interests in a space in which diverse actors operate. Creating a common framework for these different audiences will help those who operate in this domain identify similarities and synergies across types of approaches and specific projects, learn from each other, and expand possible benefits associated with energy and development efforts.

The need for coordination in this field is becoming increasingly obvious. In the United States, for example, the National Governors Association (NGA) launched a policy academy in 2013 to fund a handful of states to advance energy and economic development projects. The Center for Best Practices at the NGA recognized that state policymakers in these two fields did not work together often, simply because they never had to in the past. The NGA, therefore, saw value in pushing this coordination among energy and economic development policymakers with funded projects to help strengthen these relationships and to set an example for other states to emulate. This book has a similar goal of connecting those who operate in this domain to a common framework, but it also seeks to speak to a worldwide audience.

1.2 Reaching a Diverse Audience

The book includes detailed accounts of activities, approaches, trends, and case-specific accomplishments across different country contexts to provide examples of EBED programs through design and implementation. Thus, we blend conceptual and applied accounts of EBED to ensure its relevance to students, researchers, and practitioners. We construct the EBED framework throughout this book to help readers identify their EBED activities; set specific and deliberate goals and objectives; and, as a result, better advance EBED efforts in the future. In this spirit, we also aim for the book to initiate a broader discussion about the role of EBED activities in modern society, how they fit within the current business and policy environment, and ways to maximize their potential to confront 21st century challenges.

Given the blend of theory and application, this book has three primary intents to best serve the EBED reader. First, we *inform* students of all kinds in energy, economic development, policy, regional planning, and economics about EBED. Next we *advance knowledge* about the convergence of energy and economic development fields in the EBED domain. Third, we demonstrate EBED *applications* in a range of contexts so that practitioners can learn from examples of EBED implementation and adapt certain elements to their own EBED contexts.

1.3 Complex Challenges that Establish the Need for EBED

The need for EBED can be established on many grounds, depending on location- and context-specific factors. The kinds of issues that EBED most commonly addresses are greenhouse gases (GHGs) and other emissions, energy security and efficiency, energy poverty, economic growth and recovery, job creation, and poverty. This list is not intended to be exhaustive but rather provide an overview of energy and economic challenges that motivate EBED initiatives in different regions of the world.

1.3.1 Reduce Greenhouse Gases and Other Emissions

GHGs are a major contributor to the average warming of the climate system. The nine warmest years since 1880 occurred after 2000 (National Aeronautics and Space Administration 2013). This trend of temperature change has implications for sea-level rise, changes in agriculture and forestry, an increase in extreme weather patterns, and harmful effects on human health (Intergovernmental Panel on Climate Change [IPCC] 2007).

Heavy reliance on fossil fuel resources contributes significantly to the release of GHG emissions. In the United States, 94 % of all CO_2 emissions result from the combustion of fossil fuels (U.S. Environmental Protection Agency [EPA] 2013). Globally, 58 % of all 2007 GHG emissions came from the energy, transportation, and industrial sectors (IPCC 2007).

The continuous growth of global energy consumption is leading to an increased reliance on fossil fuel energy resources. As of 2011, 85 % of global primary energy consumption came from nonrenewable, carbon-emitting fossil fuels (U.S. Energy Information Administration [EIA] 2011). Rapidly emerging economies in particular are experiencing high rates of energy consumption growth and have become some of the world's leading GHG emitters. In 2007, energy use in non-OECD countries exceeded OECD energy use for the first time. The projected difference between non-OECD and OECD energy use by 2035 is 63 % (EIA 2010). In 2007, China and India combined accounted for 26 % of total global carbon dioxide (CO_2) emissions, up from 13 % in 1990 (EIA 2010). The projected annual increase between 2007 and 2035 in energy-related carbon dioxide emissions in OECD countries is 0.1 %, while the projected increase in non-OECD countries is 2.0 %.

The production of energy contributes to the release of other emissions as well, including sulfur dioxide, nitrogen oxides, carbon monoxide, volatile organic compounds, and particulate matter. Various combinations of these emissions, in conjunction with GHG emissions, create smog, tropospheric ozone, acid rain, and regional haze. These regional-scale pollution "soups" affect living creatures as well as biological ecosystems; some also degrade human-made surfaces (United

Nations Development Programme [UNDP] 2000). Additional pollutants from the extraction, refinement, and generation of energy contribute to water pollution. These pollutants include mercury, lead, and arsenic, as well as a number of other heavy metals.

EBED projects aim to reduce GHG and other emissions through a focus on clean and sustainable energy sources, including demand-side resources.

1.3.2 Improve Cost, Efficiency, and Energy Security

Heavy reliance on fossil fuels results in much more than pollution externalities; it can also lead to significant dependence on energy providers and render a country and its people vulnerable to fluctuations in the price and supply of energy. The price of fossil fuels is variable and, depending on the resource, may fluctuate considerably over time. Countries that rely on an energy mix with significant price volatility find it difficult to guard against price changes. Price variability and rising costs of energy resources such as fossil fuels can have significant effects on households, businesses, and governments because they can disrupt the costs of doing business, getting to and from a job, and feeding one's family. These are functions that strike at the core of society and have serious implications about how people live and work. Subsidies are often used as a means to manage this volatility. Subsidies also, however, have the potential to help lock in path dependence on specific resources, such as finite fossil fuel resources.

When entire economies rely on cheap energy or when households spend a significant proportion of their income on energy characterized by inelastic demand (i.e., when the price of a resource increases but demand for that resource does not decrease significantly), price fluctuations can lead to serious economic and social problems. Some countries such as Jordan, for example, offer significant subsidies on fuel for its citizens to guard against the negative impacts that these price fluctuations have on households, especially the large share of them at or near the poverty level. In 2013, approximately 100,000 Jordanian families received a total of JD 100 million ($141 million) in subsidies (PetroWorld 2013). The subsidy system is currently being reformed, however, to draw in international aid (Gamble 2013).

As countries guard against the disruption that can result from price volatility, these safeguards simultaneously lead to behaviors of overconsumption that are unsustainable. Unsustainable overconsumption is typically an issue when a resource—such as oil or natural gas—is finite, and the eventual exhaustion of it will require consumers to switch to alternative sources, generally at higher prices. Simple economics dictates that as the amount of a resource declines, the price of that resource will increase, assuming that technology and all other factors remain constant. Eventually the price will rise to a level that makes alternative goods cost-competitive and consumers will start to substitute one for the other. Although this transition is understandably difficult to manage, these rising costs also offer opportunities if addressed responsibly and with appropriate alternatives. Rising costs also lead to behavioral change and therein lies the opportunity for increased reliance on a sustainable energy mix. If appropriate options are in place for these households and businesses to use lower-carbon and more efficient alternatives, a more balanced energy mix can result over time.

EBED projects can help places and households adopt a more balanced and diverse energy mix by introducing new products into markets or creating jobs related to low-emissions, efficient, and advanced technology energy suppliers.

Energy efficiency is a related issue to energy security and one that presents another set of needs and opportunities for EBED. Energy efficiency refers to doing more with less, or the transition to using less energy input to achieve the same amount of energy output. Energy systems tend to have significant potential for efficiency savings on both the supply and demand sides. Energy generation typically suffers from substantial efficiency loss. A coal power plant, for example, converts an average of 33 % of primary energy into electricity; the most advanced coal technologies are able to reach efficiencies of 45 % (International Energy Agency [IEA] 2012), still leaving significant room for improvement. Transmission and distribution account for an approximately 9 % loss, on average, and can be much higher in some countries. Electricity theft also factors into these energy loss figures in several countries around the world (IEA 2010). Iraq, for example, lost approximately 37 % of its electricity production from transmission and distribution losses in 2010, down, however, from 49 % in 2008. In 2010, Haiti lost approximately 58 %, Botswana 56 %, and the Republic of Congo 83 % (World Bank 2012). Sub-Saharan Africa has the highest rate of electricity loss of any region in the world (United Nations-ENERGY/Africa 2008).

On the consumer side, residential, commercial, and industrial end users are responsible for electricity losses due to inefficient appliances and so-called "phantom loads," where an appliance such as a plasma TV or a cellular phone charger is still pulling electricity from the outlet even when it is not in the "on" position. Seven to 13 % of electricity consumption in the United States, Europe, and Japan is attributable to residential phantom loads (*The Economist* 2006).

Efficiency losses are not exclusive to electricity sectors. The transportation sector also is associated with significant energy losses. For example, only 14–26 % of the fuel used by a standard light-duty automobile in the United States is converted to useful energy (U.S. Department of Energy [DOE], Office of Energy Efficiency & Renewable Energy and EPA, Office of Transportation and Air Quality 2013).

Unreliable and inefficient energy can impede industrial production, economic growth, and economic opportunity. One source estimates that unreliable and decaying energy infrastructure costs Sub-Saharan Africa approximately one-

---

[1] Undercollection is when a utility company is not able to collect payments for the electricity consumed by its end users, either because it does not have mechanisms in place to collect payments or to track exactly how much is used by each consumer or because end users steal electricity.

quarter of a percentage point off of its annual gross domestic product (GDP) growth rate per year. The combination of utility undercollection[1] and efficiency losses amounts to approximately $2.7 billion per year, or 0.8 % of GDP on average (Eberhard et al. 2009) and upwards of 1–2 % of annual growth potential in Sub-Saharan Africa (Foster and Briceño-Garmendia 2009).

EBED programs incorporate energy-efficient techniques in ways to reduce costs, encourage business and industry development, and spur job creation.

1.3.3 Reduce Energy Poverty

Over 25 % of the world's population, between 1.4 and 1.6 billion people, live without electricity. Approximately 3 billion people rely on traditional biomass for cooking and heating (UNDP and World Health Organization 2009). Insufficient energy access, which is defined as "access to clean, reliable and affordable energy services for cooking and heating, lighting, communications and productive uses" (Advisory Group on Energy and Climate Change 2010, p. 13), is prevalent in least developed countries, but it also affects low-income individuals in all countries. Despite the increased focus of energy access on the international development agenda and the global awareness of the issue of energy poverty over the past two decades, the need for basic energy access in some regions like Sub-Saharan Africa has not changed much over this time, nor is it projected to change significantly over the next two decades in light of population growth (Bazilian et al. 2010).

Energy access is crucial for creating economic opportunity and fostering development and growth. Energy is a basic, enabling good that helps individuals and nations achieve other development outcomes, such as those highlighted in the Development Millennium Goals (Brew-Hammond and Kemausuor 2009; Goldemberg and Johansson 2004). Without access to electricity, it is difficult to ensure adequate health or educational facilities, sanitation, food, or water. School-aged children, for example, are limited to studying during daylight hours without energy and hospitals cannot refrigerate vaccines without power. A lack of energy access also limits the economic opportunities available to households and small businesses. Reliance on traditional biomass—such as wood, dung, charcoal, or coal—for cooking and heating can produce serious health effects for those regularly exposed to the smoke, such as young children, and can also exacerbate gender inequality when women are the primary gatherers of the fuel. Energy is also necessary for economic growth, especially at lower-income and lower-middle-income country levels, whereas at higher country income levels the relationship between energy access and economic growth is bidirectional (Apergis and Payne 2011).

1.3.4 Identify Energy-Based Economic and Employment Opportunities

Most, if not all, places strive to develop in a way that generates economic opportunities and wealth creation for its residents. Some regions are underperforming and economically depressed and seeking ways to generate jobs and income. Others are emerging in industry and economic growth and aim to find a path to a healthy and dynamic economy that can support critical investments in modern physical infrastructure (e.g., highways, electricity, Internet access) and soft infrastructure (e.g., education and healthcare). Finally, advanced economies typically strive to "stay on top" of global economic competitiveness to ensure continued economic prosperity. Regardless of the position a country or locale is in, these places hold a common desire to grow or maintain an economy that can support them.

Recently, because of the events of the 2007 to 2009 recession experienced by many countries, the desire for economic opportunity was coupled with acute economic needs. Worldwide output in several sectors, along with trade and stock market indicators, fell farther during the recession than during the Great Depression in the 1930s (Almunia et al. 2010). The global economic downturn of 2007 led to a massive number of layoffs across the world. Several years post-recession, unemployment rates still remain quite high and, in some countries, continue to rise, such as South Africa, Spain, Italy, and the United Kingdom. Unemployment rates in 2011 ranged from approximately 3.5 % in South Korea and 4 % in Japan to over 25 % in South Africa and 22 % in Spain (U.S. Bureau of Labor Statistics 2012).[2]

As countries sought to dig their way out of the recession, many recognized that the energy sector and related sectors within typical energy supply chains offer opportunities for employment growth. The United Nations Environmental Programme (UNEP), for example, argues that "the greening of economies is not generally a drag on growth but rather a new engine of growth; that it is a net generator of decent jobs, and that it is also a vital strategy for the elimination of persistent poverty" (UNEP 2011, p. 10). The large number of stimulus packages that provided funding for energy and "green" industries across the world between 2008 and 2011 also underscores the investments countries were willing to make for the potential to create jobs from energy-related development.

1.3.5 Identify Energy Links to Alleviate Poverty

Chronic and persistent poverty remain a global problem regardless of the energy or economic composition of a country. Approximately 2.4 billion people around the

---

[2] These data are gathered by the U.S. Bureau of Labor Statistics. The data only include 16 countries. Unemployment estimates are generated using U.S. employment concepts.

world live in poverty. Roughly half, or 1.2 billion, live in extreme poverty. Poverty is defined as "pronounced deprivation of well-being" (Soubbotina 2004, p. 33). In consideration of Amartya Sen's (1987) conceptualization of well-being as the capacity to function in society, Haughton and Khandker (2009, pp. 2–3) define poverty as a case "when people lack key capabilities," such as adequate income, health, education, rights, and opportunities. Extreme, or absolute, poverty refers to severe deprivation of basic human needs, such as water, food, sanitation, shelter, health, and education. The World Bank measures poverty as living on US$2.00 per day or less, using purchasing power parity. Extreme poverty is measured as living on less than $1.25.

Rates of poverty have declined over the past several decades. In 1990, approximately 43.1 % of the world lived on $1.25 per day or less; by 2010, 20.6 % of the world's population lived at this level of extreme poverty (World Bank 2013). Much of this improvement is attributable to changes in East and South Asia. The poverty rate in Sub-Saharan Africa, however, has changed very little over this time. Poverty, of course, also exists in developed economies such as the United States and within Europe. The poverty rate in the United States in 2011 was 15.0 %, representing 46.2 million people (DeNavas-Walt et al. 2012). In the European Union, 16.9 % of the population was at risk of poverty after social transfers in 2011 (Eurostat 2013).

Energy and climate problems exacerbate poverty. It is generally expected that climate change will affect developing countries the hardest, where the majority of the world's poor reside. The geographic location of many countries along the equator, surrounded by ocean, or in the line of specific weather patterns puts them at greater risk of sea level rise, changes in ocean currents, changes in precipitation patterns, increased temperatures, and increased weather intensity. These impacts on the natural environment will affect agricultural outputs, availability of clean water, and the prevalence of disease, all of which will make a transition out of poverty all the more challenging to overcome.

The EBED framework helps shape efforts to link energy production and deployment to approaches that help reduce poverty.

1.4 Outline of the Book

We introduce EBED to the reader in two parts. Part I informs and advances knowledge about EBED. Part I lays the conceptual framework for EBED and addresses questions such as what is EBED, which activities are included in the EBED framework, how in general terms is EBED implemented, which policies support EBED activities, and how can one evaluate EBED initiatives? Part II is more applied and uses a series of case studies that portray nationally and locally driven EBED approaches. We illustrate how the EBED framework relates in these cases and discuss considerations for replication and scale for future EBED efforts.

1.4 Outline of the Book

The book concludes with an analysis of these applications and broader EBED trends.

Part I begins with Chap. 2, in which we define the domain of EBED. This chapter first provides a definition of EBED projects and then outlines the common energy and economic development goals. It discusses the foundations of the discipline, with a detailed account of how activities in the energy policy and planning and economic development fields have evolved through the years to converge at the intersection of EBED.

Chapter 3 presents an EBED process that helps program designers and implementers identify specific goals and objectives and shape, execute, and evaluate their efforts accordingly. This chapter proceeds step-by-step through this EBED process. We describe the different stakeholders involved, the data that need to be collected, and the importance of establishing metrics and evaluation mechanisms as part of the process. We also discuss three additional factors for consideration when designing and implementing an EBED approach: the point of intervention, the geographic scope, and the scale of transformation. Point of intervention includes place-based, market-based, and household-based EBED approaches. Geographic scope is divided into top-down and bottom-up efforts, which is a theme that we carry over into the case study chapters as well. Scale of transformation refers to the degree to which new initiatives represent concrete breaks from past trends or incremental changes on business as usual. All of these considerations help EBED proponents make thoughtful and deliberate choices about how to proceed with their work.

In Chap. 4, we discuss various policies that can be used to help achieve EBED outcomes. We describe seven policies that enable, accelerate, or regulate EBED activities:

- technological innovation policies
- technological adoption policies
- entrepreneurship policies
- industrial growth policies
- workforce development policies
- climate and environmental policies
- planning.

For each policy type, we provide specific examples and explain how these categories are related to the various EBED approaches presented in Chap. 3.

Chapter 5, which focuses on EBED evaluation and metrics, concludes the first part of the book. This chapter explores evaluations of EBED projects to date, with a consideration of the outcome measures most commonly reported, the methods used, the timing of evaluation, and the type of initiative most commonly analyzed. This chapter concludes with a set of recommendations for evaluators to help them tailor evaluations to project goals and conditions.

In Part II of the book, we demonstrate how EBED is applied. Chapters. 6 through 9 present case studies of current or recent EBED initiatives. We divide this discussion according to the level of implementation of EBED projects and begin with locally driven EBED approaches and then turn to nationally oriented efforts that involve more "whole-of-government" approaches. The case studies that we highlight represent a range of different program elements, including energy, economic, environmental, and social dimensions.

Chapter 6 introduces the case studies and Chap. 7 presents local approaches. These bottom-up initiatives originate and are implemented at the subnational level by state or local governments, nonprofits, private–public partnerships, firms, or individuals. The local case studies we review are Nuru Energy in Rwanda, Kamworks in Cambodia, natural gas development in southwestern Pennsylvania, the Bandeirantes Landfill Gas to Energy Project in Brazil, Clean Energy Works in Oregon, and the Copenhagen Cleantech Cluster in Denmark.

Chapter 8 presents nationally driven initiatives that are designed and administered by federal governments. As our case studies reveal, most top-down initiatives involve a great deal of subnational participation but the funding, direction, or oversight of the initiative is primarily handled by the national government. The top-down case studies we review are the national hydropower initiative in Laos, the "NexBTL" foreign direct investment project in Singapore, the China Golden Sun initiative in China, Ethiopia's national cookstoves program, the Morocco Solar Initiative, and the South Africa Renewables Initiative (SARi).

We devote Chap. 9 to an overview of the U.S. ARRA, an example of a hybrid top-down and bottom-up approach that involved substantial funding and a diverse range of EBED projects and has contributed valuable insights regarding the timing, effectiveness, and feasibility of EBED projects.

Chapter 10 provides a discussion of trends and implications revealed in the case studies. We also summarize the main takeaways about the EBED framework presented throughout the book.

References

Advisory Group on Energy and Climate Change (AGECC) (2010) Energy for a sustainable future: report and recommendations. New York

Almunia M, Bénétrix A, Eichengreen B, O'Rourke KH, Rua G (2010) From great depression to great credit crisis: similarities, differences and lessons. Econ Policy 25:219–265

Apergis N, Payne JE (2011) A dynamic panel study of economic development and the electricity consumption-growth nexus. Energy Econ 33:770–781

Bazilian M, Sagar A, Detchon R, Yumkella K (2010) More heat and light. Energy Policy 38:5409–5412

Brew-Hammond A, Kemausuor F (2009) Energy for all in Africa—to be or not to be?! Curr Opin Environ Sustain 1:83–88

DeNavas-Walt C, Proctor BD, Smith JC (2012) Income, poverty, and health insurance coverage in the United States: 2011. U.S. Census Bureau. http://www.census.gov/prod/2012pubs/p60-243.pdf. Accessed 21 June 2013

Eberhard A, Foster V, Briceno-Garmendia C, Shkaratan M (2009) Power: catching up. In: Foster V, Briceño-Garmendia C (eds) Africa's infrastructure: a time for transformation: the

References

International Bank for Reconstruction and Development/The World Bank, Washington, DC, pp 181–202

Eurostat (2013) Income distribution statistics—statistics explained. http://epp.eurostat.ec.europa.eu/statistics_explained/index.php/Income_distribution_statistics#At-risk-of-poverty_rate_and_threshold. Accessed 21 June 2013

Foster V, Briceño-Garmendia C (2009) The Africa infrastructure country diagnostic. In: Foster V, Briceño-Garmendia C (eds) Africa's infrastructure: a time for transformation. The International Bank for Reconstruction and Development/The World Bank, Washington, DC, pp 31–42

Frankfurt School—UNEP Collaborating Centre for Climate & Sustainable Energy Finance (2011) Global trends in renewable energy investment 2011: analysis of trends and issues in the financing of renewable energy. Frankfurt School—the UNEP Collaborating Centre for Climate & Sustainable Energy Finance, Frankfurt School of Finance & Management, Frankfurt/Main

Gamble LH (2013) Jordan asks for international aid to deal with Syrian refugees. CNBC, May 28. http://www.cnbc.com/id/100768665. 21 June 2013

Goldemberg J, Johansson TB (eds) (2004) World energy assessment: overview update 2004. United Nations Development Programme, New York

Haughton J, Khandker SR (2009) Measuring and analyzing poverty: a handbook on poverty and inequality. The World Bank, Washington, DC

Intergovernmental Panel on Climate Change (IPCC) (2007) Climate change 2007: synthesis report. http://www.ipcc.ch/publications_and_data/ar4/syr/en/spm.html. Accessed 29 January 2013

International Energy Agency (IEA) (2012) Technology roadmap: high-efficiency, low-emissions coal-fired power generation. Overview. http://www.iea.org/publications/freepublications/publication/name,32869,en.html. Accessed 21 October 2013

International Energy Agency (IEA) (2010) Energy technology perspectives 2010: scenarios and strategies to 2050. IEA Publications, Paris

Kirkman GA, Seres S, Haites E, Spalding-Fecher R (2012) Benefits of the clean development mechanism 2012. United Nations Framework Convention on Climate Change. United Nations Climate Change Secretariat, Bonn

National Aeronautics and Space Administration (2013) NASA finds 2012 sustained long-term climate warming trend. http://www.nasa.gov/topics/earth/features/2012-temps.html. Accessed 23 June 2013

Organisation for Economic Co-operation and Development (OECD) (2013) Query wizard for international development statistics. http://stats.oecd.org/qwids/. Accessed 15 May 2013

PetroWorld (2013) Jordan: government releases new subsidy payments. PetroWorld, April 17. http://www.petrolworld.com/africa-middle-east-headlines/jordan-government-releases-new-subsidy-payments.html. Accessed 21 June 2013

Rittel HWJ, Webber MM (1973) Dilemmas in a general theory of planning. Policy Sci 4:155–169

Sen A (1987) Commodities and capabilities. North-Holland, Amsterdam

Soubbotina TP (2004) Poverty and hunger. In: Soubbotina TP (ed) Beyond economic growth: an introduction to sustainable development, 2nd edn. The International Bank for Reconstruction and Development/The World Bank, Washington, DC, pp 33–42

The Economist (2006) Pulling the plug on standby power. The Economist, March 9. http://www.economist.com/node/5571582. Accessed 21 October 2013

U.S. Bureau of Labor Statistics (2012) International comparisons of annual labor force statistics. http://www.bls.gov/fls/flscomparelf/lfcompendium.pdf. Accessed 1 February 2013

U.S. Department of Energy (DOE), Office of Energy Efficiency & Renewable Energy and U.S. Environmental Protection Agency, Office of Transportation and Air Quality (2013) Fuel economy: where the energy goes. www.fueleconomy.gov. http://www.fueleconomy.gov/feg/atv.shtml. Accessed 21 October 2013

U.S. Energy Information Administration (2011) International energy outlook 2011.DOE/EIA-0484A(2011). Washington, DC

U.S. Environmental Protection Agency (EPA) (2013) Inventory of U.S. greenhouse gas emissions and sinks: 1990–2011. http://www.epa.gov/climatechange/Downloads/ghgemissions/US-GHG-Inventory-2013-Main-Text.pdf. Accessed 21 October 2013

United Nations Development Programme (UNDP) (2000) World energy assessment: energy and the challenge of sustainability. http://www.undp.org/content/undp/en/home/librarypage/environment-energy/sustainable_energy/world_energy_assessmentenergyandthechallengeofsustainability.html. Accessed 12 March 2012

United Nations Development Programme (UNDP) and World Health Organization (2009) The energy access situation in developing countries—a review focusing on least developed countries and Sub-Saharan Africa. Sustainable Energy Programme Environment and Energy Group Report. United Nations Development Programme, Environment and Energy Group, Bureau for Development Policy, New York

United Nations Environment Programme (UNEP) (2011) Towards a green economy: pathways to sustainable development and poverty eradication—a synthesis for policy makers. United Nations Environment Programme, Geneva

United Nations-ENERGY/Africa (2008) Energy for sustainable development: policy options for Africa. United Nations, New York

World Bank (2012) Electric power transmission and distribution losses (% of output). http://data.worldbank.org/indicator/EG.ELC.LOSS.ZS. Accessed 29 January 2013

World Bank (2013) Poverty and equity data. http://povertydata.worldbank.org/poverty/home/. Accessed 15 May 2013

Chapter 2
Defining Energy-Based Economic Development

Abstract EBED is a direct extension of the energy planning and economic development disciplines. This chapter provides important foundational information to support the EBED framework and applications described in the remainder of the book. It defines EBED, describes the economic development and energy policy and planning disciplines, summarizes how these disciplines have converged, and discusses the goals inherent within EBED initiatives. This chapter also differentiates EBED from other disciplines. Distinguishing characteristics include EBED's focus on advanced, efficient, or low-emissions energy; its pursuit of joint energy and economic development goals; ability to build on the varying scale and distributed nature of some types of low-emissions energy; its alignment of development and energy goals into one unified approach; and the active role of governance, leadership, and stakeholder engagement.

2.1 Definition

EBED includes activities at the nexus of energy policy and planning and economic development disciplines, where practitioners seek to address the challenges and opportunities within the juncture of these two fields. This domain embraces initiatives, programs, or policies that share the goals of economic development and energy. Several important elements within this conceptualization of the EBED domain distinguish its practices from other efforts. In particular, EBED

- focuses on advanced, efficient, or low-emissions energy sources and technologies;
- advances joint energy and economic development goals;
- builds on the varying scale and distributed nature of low-emissions energy;
- provides a framework that aligns goals in a unified approach; and
- recognizes the role of governance, leadership, and stakeholder models in shaping outcomes.

EBED focuses on energy sources that are advanced, efficient, or low-emissions[1] and that are driving new forms of economic development (Carley et al. 2011, 2012). Advanced means innovative technological improvements on conventional or alternative energy. Efficient technologies use less input energy to produce the same amount of output energy. A low-emissions technology produces limited amounts of emissions throughout its entire production cycle.

First, many researchers and practitioners highlight renewable energy and energy efficiency as promising solutions to our current global and local energy challenges. In this book, we highlight and discuss renewable energy and energy efficiency innovations and examples of how places around the world are adopting these innovations for diversification or decarbonization reasons. We recognize, however, that renewable energy and energy efficiency will comprise only a portion of our future energy scenario and are also more relevant in some regions than in others. Extractive industries such as coal, gas, oil, and rare elements, for example, are also relevant to these discussions if they are extracted in ways that are more advanced, produce less emissions, and are more efficient. For these reasons, researchers and practitioners working in a range of energy fields are important to the EBED discussion.

Second, EBED initiatives, by their definition, advance both economic development and energy goals. Economic development is a significant element in the adoption of energy solutions at national and subnational levels and is also important in the mitigation of energy-related problems, such as the quest for reducing GHG emissions. Energy is an important aspect of economic development because, as an industry, the energy sector is in a state of significant transformation with increased global demand and limited supply of natural resources. As an ever-present and required input for growth, energy creates an opportunity for innovation to meet the demand for new kinds of energy provision. When there is need for innovation there is typically opportunity for market-based growth, thus making the energy domain a hot bed for economic development potential. In short, energy has become a driver of economic development, not just an enabler of it.

Third, EBED is conducive to varying scales and distribution models that can be widely deployed. The assets on which EBED initiatives can build differ widely from location to location, but the more distributed nature of lower-carbon energy sources widens the potential for the kinds of markets, places, and households that can experience the economic benefits from EBED. We showcase in this book the diverse projects and places that are taking advantage of EBED by leveraging their unique energy and economic development asset base.

Fourth, EBED initiatives use a framework—presented in Chap. 3—that aligns goals, objectives, outcomes, and measurement to ensure project outcomes. This

---

[1] The original conceptualization of EBED focused on "low-carbon," rather than "low-emissions" energy solutions. In this book, we expand the definition from low carbon to low emissions to highlight the importance of technologies and other solutions that not only aim to reduce GHG emissions, but also seek low levels of other pollutants as well, including but not limited to SO_2, NO_x, and particulate matter.

2.1 Definition

alignment of the various components of the EBED process helps involve a wider set of champions to advance common causes. As one scholar articulates, "Starting with a green growth and development goal is far more than a rhetorical device in that it engages different political and economic actors, and can achieve more effective progress by aligning their plans, interests and leadership" (Zadek 2011, p. 1062).

Finally, and closely related to the last point, the EBED framework recognizes the role of governance, leadership, and stakeholder engagement in shaping desired outcomes. Incorporating the role of actors and institutions within the EBED process accounts for political context and human realities that are often difficult to quantify but can have significant effects for policy and program success. Späth and Rohracher (2010) underscore this point in their work on energy regions by stating:

> The participation of a specific variety of actors from different parts of society can be of paramount importance for the momentum that such an alliance can generate. Only an alliance including major businesses, NGOs and government authorities can assume enough authority for their joint objectives, agenda and standards because a more homogenous or incomplete constellation would not be trusted to produce socially robust and operational solutions…but rather partial interests (p. 454 and referencing Boström 2006).

This point is important because EBED is driven by a diverse set of actors. The framework EBED provides helps these actors and the institutions they represent work from a common understanding that incorporates local context. Energy policy and planning decisions are typically made solely by utility companies, regulators, and policymakers. The EBED realm incorporates actors such as economic developers, government officials, civil servants, the community, community development practitioners, donors, industry and business leaders, and representatives of the nonprofit realm. This integration of diverse actors in EBED activities builds a more robust network for cross-fertilizing activities. As a result, there is greater potential for more innovative and prolific resolution of EBED issues.

2.1.1 Other Definitions of the Energy–Development Nexus

Others have used several working definitions to describe the juncture of energy and economic development. Terms such as "green economy" and "green jobs" are sometimes used, but these definitions can be easily applied for advocacy and political purposes and thus obfuscate the intent of the terms and make it more challenging for different audiences to find a common understanding. For example, "green" can mean a variety of things to different people or organizations (Fitzgerald 2010; Peters et al. 2010). We seek to be more explicit in the definition of EBED to enhance communication among practitioners from a variety of backgrounds. Clearly defined terms facilitate the development of cohesive objectives and performance metrics and make it easier to design, evaluate, and scale successful programs.

Other definitions are more specific. Nissing and von Blottnitz (2010) use a term "sustainable energisation," for example, to describe this nexus of energy and economic development. This definition is constructive because it focuses on different levels of energy development, as defined according to primary and secondary energy service needs. Primary energy service needs are defined as entirely consumptive in nature. They do not make significant contributions to financial or economic development but enable short-term poverty alleviation and improved quality of life. Conversely, secondary energy service needs contribute to productive output, such as income growth, skills development, and infrastructure development, which in turn leads to medium- to long-term economic development (Nissing and von Blottnitz 2010). Sustainable energisation is defined as the following:

> The transitional process of progressively meeting primary and early secondary energy service needs of a poor economic subgroup (second economy) through the delivery of an enhanced quantity, quality and/or variety of accessible and affordable energy services, enabling the sustainable development of the considered subgroup based on poverty alleviation and economic development, as well as the optimization of the energy service supply network from a lifecycle perspective. (Nissing and von Blottnitz 2010, p. 2186)

The distinction that Nissing and von Blottnitz make between primary and secondary energy service needs is important. "Energisation" can be applied to alleviate poverty by supplying energy services and in ways that boost growth or development. The core concept of "energisation," however, is the transition from energy supply (i.e., primary needs) to growth and development (i.e., secondary needs) in a developing-country context. Nissing and von Blottnitz highlight three important sustainable energisation targets in their work:

- poverty alleviation
- transformative strategies that provide a bridge between poverty alleviation and economic growth
- efforts directed toward boosting economic growth and development.

Although we believe "energisation" provides a constructive overview of activities in the energy and development nexus, it excludes a range of EBED activities. EBED is broader than sustainable energisation in that it includes efforts to improve access to and availability of energy (i.e., primary) and ways to supply energy to promote long-term development (i.e., secondary). EBED, for example, can include research and development (R&D) efforts to create a new technology in support of energy-efficient lighting or workforce training programs that teach energy retrofitting skills in construction. These examples do not promote access or long-term economic development; rather, encourage employment, affordable energy use, and efficient energy use.

Later in this chapter, we expand on these points by presenting the spectrum of energy and economic development goals EBED advances. First, we delve more deeply into the foundation of the energy and economic development disciplines to describe the underpinnings of the EBED framework.

2.2 Foundations

Understanding the basis for the convergence of energy planning and economic development advances knowledge about the EBED domain and its evolution. This overview of the founding disciplines is intended to help the student, researcher, and practitioner better recognize the strategic use of elements from one or the other discipline to improve EBED outcomes.

2.2.1 Economic Development

Economic development is a dynamic field that is defined in many different ways, depending on one's vantage point. We define it simply as "a process of creating wealth for regions and improving the economic opportunities for the people that live and work within them. Desired results from this process include improved standards of living and reduced levels of poverty" (Carley et al. 2011, p. 283). We offer additional definitions to highlight different perspectives. Malizia (1994) defines economic development as

> The ongoing process of creating wealth in which producers deploy scarce human, financial, capital, physical and natural resources to produce goods and services that consumers want and are willing to pay for. The economic developer's role is to participate in the process of national wealth creation for the benefit of local consumers and producers by facilitating either the expansion of job opportunities and tax base or the efficient redeployment of local resources (pp. 83–84).

This definition underscores the importance of wealth creation, production systems, and the role of the developer in facilitating the development process for the benefit of those who live and work in the location where the development occurs. Bolton (1991) emphasizes economic distress in his definition, and Eisinger (1998) extends this perspective to specify that the ultimate intent of economic development is to "enhance the collective well-being" of communities (p. 6). Amsden sees economic development as a process from moving from an economy based on primary products that uses unskilled labor to an economic asset base of knowledge that requires skilled labor (2001, p. 2).

Regardless of the vantage point from which economic developers approach their work, ultimately they try to create economic opportunities for businesses (e.g., increasing profitability) and workers (e.g., higher skilled and higher paid jobs) and to spur growth and development for specific locations. Some examples of interventions that may be tailored to specific locales include industrial growth strategies, industry recruitment and retention, export promotion, business investment, microenterprise development, special economic zones, entrepreneurship training, and worker training.

Two features of economic development are particularly important in the context of EBED. First, the increased importance of innovation and knowledge as a driver of economic development and the transformation of economies has elevated the role of education, R&D, and entrepreneurship in development. Colleges, universities, and R&D centers tend to have more prominent and active roles in economic development than they did in decades past. For example, countries are now designing universities to focus heavily on commercialization, not just education. King Abdullah University of Science and Technology in Saudi Arabia is one such example: it focuses on creating a "knowledge-based economy" through research, product development and transfer, entrepreneurship, industry collaboration, and policy (King Abdullah University of Science and Technology 2013).

Second, sustainability and the "triple bottom line" of coordinated economic, environmental, and social development are increasingly demanded in economic development efforts (Stimson et al. 2006). For decades, environmental consequences of growth remained tangential to economic development because mitigating against them was commonly viewed as an inhibitor to growth. In fact, proponents of sustainability and economic development were often at odds with each other. This is becoming less prevalent because of a steadily growing awareness about the value of a clean environment to economic development and the direct costs of heavy pollution. For example, rapid industrialization in China since the late 1970s has resulted in economic growth that is "nothing short of spectacular" and contributing "considerable poverty reduction in its wake" (Hausmann et al. 2008, p. 356 cited in Serra and Stiglitz 2008). One of the main costs of this growth is ambient air pollution and the significant impact of this air pollution on public health (Kan et al. 2012). Ambient air pollution, as of 2010, was in the top four mortality risk factors for China and in the top seven worldwide. In China, air pollution contributed to approximately 3.2 million deaths in 2010 (Wong 2013). In response, the Chinese government has established targets to limit air pollution in Bejing, Tianjin, and Hebei, cities with the most extreme smog (China Daily USA 2013). The growing recognition that there are real costs to human health and worker productivity is forcing sustainability into the economic development discipline more than ever.

2.2.2 Energy Policy and Planning

The field of energy policy and planning includes actions taken by government, not-for-profits, or private organizations to plan energy resource use, develop energy technologies, and develop policy instruments and regulations to shape direct energy (i.e., heat) or secondary energy (i.e., electricity) production and consumption. These efforts encompass the full fuel cycle of all energy resources—location, extraction, transportation, refinement, processing, distribution, combustion or other use, and waste disposal. It also includes the supply-side resources and technologies used to produce energy, factors and approaches related to energy consumption such as energy efficiency and load control, and information or knowledge about energy resource management.

2.2 Foundations

The coalescence of some of the issues described in Chap. 1, including the increasing recognition of climate change and its negative effects, volatile energy prices, and an increasing focus on energy security and energy access, has raised awareness of the importance of energy policy and planning to preserve economic stability. The potential for massive blackouts (such as those occurring in Northern India in 2012 and California, USA, in 2000 to 2001) due to unreliable infrastructure and the risks that these electrical malfunctions pose to businesses reaffirm the significant connection, if not reciprocal relationship, between energy development and economic growth. Energy policy and planning is also becoming increasingly common not merely as a response to environmental, social, or economic problems, but as an opportunistic market-based approach because energy technologies have significant potential for future marketability.

Like economic development, the shape and form of energy policy and planning depend on the country context. For example, Germany embraced a progressive energy policy platform beginning in the 1970s to improve its energy security, though this evolved into an environmental platform in the 1990s, as well as a later goal of replacing nuclear generation (Runci 2005; IEA 2013a). In contrast, in countries like the United States that have less prominent national leadership on energy and climate policy, besides electricity and natural gas market oversight and alternative energy tax incentives, many state and local entities—both governmental and nongovernmental—have stepped into leadership roles. These entities have initiated efforts to increase diversification of energy sources, increase energy self-sufficiency, or both. Instead of focusing on carbon mitigation policies, many of this energy policy and planning is motivated by economic development objectives in pursuit of "home-grown" energy or as a means of diversification of state or regional economies to improve competitiveness (Rabe 2004, 2008). Many recent state and local energy strategies attempt to gain an early alternative energy market share and profit from future energy developments.

Also similar to economic development, energy policy and planning efforts have become increasingly focused on energy technology innovation, which is defined by Gallagher and her colleagues (2006) as

> The set of processes by which improvements in energy technology, which may take the form of refinements of previously existing technologies or their replacement by substantially different ones, are conceived; studied; built, demonstrated, and refined in environments from the laboratory to the commercial marketplace; and propagated into widespread use (p. 195).

Energy technology innovation focuses on development and deployment of efficient, reliable, advanced, and low- to no-carbon energy technologies, including demand- and supply-side technologies. These technologies are intended to serve one of two roles: (1) replace or enhance conventional energy sources or (2) bypass technologies based on conventional sources to advance technological solutions beyond what is in existence.

Some regions have emerged as "first movers" in terms of energy innovation by testing energy policies and programs that thrust the emerging research and development into new and highly anticipated growing markets. For example, China has invested significant resources since the early 2000s into their solar technology market, in an effort to develop manufacturers of solar technologies—such as solar photovoltaic panels—that are leaders in global markets. Over the past several years, China has invested more in the renewable energy industry than any other country (Frankfurt School—UNEP Collaborating Centre for Climate & Sustainable Energy Finance 2012). These investments, particularly in the solar industry, have led China to hold a dominant market position, relative to others, in solar panel, wind turbine, and other energy technology markets (Bradsher 2010).

2.2.3 Convergence

As these fields continue to evolve, their convergence is becoming more prevalent because of the shift of energy from functioning merely as an enabler to functioning as a driver of economic development. The two disciplines, of course, have shared fundamental connections since the beginning of human and economic development. These linkages are best described through the lens of an "energy ladder" (Barnes and Floor 1996). Toman and Jemelkova (2003) summarize this concept in the following passage:

> ... linkages among energy, other inputs and economic activity clearly change significantly as an economy moves through different stages of development ... at the lowest levels of income and social development, energy tends to come from harvested or scavenged biological sources (wood, dung, sunshine for drying) and human effort (also biologically powered). More processed biofuels (charcoal), animal power, and some commercial fossil energy become more prominent in the intermediate stages. Commercial fossil fuels and ultimately electricity become predominant in the most advanced stages of industrialization and development ... energy resources of different levels of development may be used concurrently at any given stage of economic development: electric lighting may be used concurrently with biomass cooking fires. Changes in relative opportunity costs as well as incomes can move households and other energy users up and down the ladder for different energy-related services (p. 3).

The most basic connection point between energy access, human development, and economic well-being is that energy is a fundamental enabler of basic human needs, such as food, shelter, clean water, transportation, healthcare, and education. Although this the relationship has always existed, in the last two decades the global understanding of the importance of energy as it relates to basic human needs has increased. Accordingly, international donors have intensified their investments at the connection points between energy and human development (UNDP 2005; Nissing and Blottnitz 2010; World Bank 1996).

Energy is also a primary factor of production and a crucial component of every business and economy. Our modern economy rests on the availability of energy—energy for electricity, transportation, production processes, and a variety of other economic needs.

Economic development and energy policy and planning are accelerating with new emphasis because the need for solutions to interrelated social, economic, and environmental problems is becoming more pronounced. It is now becoming common to see economic development initiatives that comprise energy strategies or vice versa. The evolution and current trends in each field demonstrate synergies that support the development and progress in the other and, thus, a greater integration of the two.

In addition to parallels in how the fields have emerged, the goals of each discipline frequently complement those of the other. Energy policy and planning seeks to improve energy self-sufficiency, which can result in the creation of businesses that are unlikely to relocate outside a given region, and increase energy diversification, which can result in the creation of new technologies, businesses, and jobs. Economic development initiatives may seek to catalyze growth through innovation, which can result in increases in energy efficiency or the creation of new technologies that diversify a given region's energy sources.

2.3 EBED Goals

EBED activities are driven by simultaneous pursuits of energy and economic goals. We identify ten EBED goals in Table 2.1 and discuss each in turn in the sections that follow. Five goals are under the realm of energy policy and planning and the other five goals under the realm of economic development. Aligning a set of goals from each realm under one framework helps one shape programs, track progress, and report successes and failures. We also reiterate here that what distinguishes EBED from other activities is that EBED efforts always involve the combination of some subset of energy and economic goals.

2.3.1 Energy Goals

2.3.1.1 Energy Diversification

Energy diversification refers to the mix of energy sources that comprises a region's energy portfolio. The concept generally implies a transition away from heavy reliance on one type of energy toward a blend that includes a variety of energy resources. Although it is not always the case that a greater diversity of energy sources is better, it is often the case that exclusive reliance on one energy type is risky. Exclusive or nearly exclusive reliance makes a region particularly vulnerable to price volatility, security threats, weather abnormalities, and other supply disruptions. For example, diversification efforts may target centralized power applications, an increase in the number of and reliance on natural gas power plants, wind turbines, or concentrated

Table 2.1 EBED goals by discipline

Energy policy and planning	Economic development
Diversify energy sources	Drive industry growth
Reduce greenhouse gas emissions and related environmental impacts	Increase innovation and entrepreneurship
Increase energy efficiency	Increase regional income
Increase energy security	Decrease poverty
Reduce energy poverty	Create jobs

solar thermal facilities. Alternatively, diversification efforts may focus on decentralized power applications, for which a diverse energy mix may include distributed generation units such as solar photovoltaic panels, nuclear modules, natural gas micro-turbines, or combined heat and power systems.

Denmark offers an example of a country that set clear goals to diversify its energy base after the oil shocks of the 1970s. At that time, Danish policymakers and citizens realized that the country was heavily reliant on outside sources for energy and was, therefore, highly vulnerable to fuel shocks. The Danish were also becoming increasingly worried about the GHG emissions associated with their heavy dependence on foreign fossil fuels. As a result, the Danish government set a goal to operate entirely off renewable sources by 2050. It expects to be one-third of the way there by 2020 (Moss 2012). The Danish have diversified their energy mix through an increase in large-scale wind energy and biomass plants, coupled with energy efficiency measures, behavior measures, and a push for electric vehicles in the transportation sector.

2.3.1.2 Reduce GHG Emissions and Related Environmental Impacts

As discussed in Chap. 1, GHG emissions from energy use are a major contributor to the average warming of the climate system and a major contributor to other emissions to air and water pollution. Reducing emissions from electricity, industry, transportation, and household use in both developed and developing countries is a critical goal for energy policymakers and planners. Increasingly, the implications of warming global temperatures are being examined across disciplines such as health, economics, and governance so that planning efforts can accommodate anticipated changes resulting from climate change. For example, as air quality diminishes, especially in urban areas, efforts to mitigate and adapt to the pollution are becoming standard local government functions.

Cities and countries alike are embracing EBED policies and programs that aim to reduce emissions, particularly GHG emissions but also including other forms such as sulfur dioxide, nitrogen oxides, particulate matter, and mercury. Examples include the national government in South Africa through its SARi and the European Union's commitment to reduce its GHG emissions by 20 % of 1990 levels by 2020 through an emissions trading scheme, renewable energy standards, energy efficiency measures, carbon capture, and reduced emissions from vehicles (European Commission 2013). As yet another example, Jakarta built a bus rapid transport system that relies on a combination of diesel and compressed natural gas (CNG) to service an estimated 39 million riders (C40 Cities n.d.). This transport system has the potential to increase mass transit and reduce CO_2 emissions by replacing petroleum with CNG.

2.3.1.3 Energy Efficiency

The goal of energy efficiency is to use less energy to achieve the same purpose. Energy efficiency efforts generally encourage energy consumers (i.e., households, businesses, and governments) to use less energy to receive the same level of energy output. Energy efficiency, which fits under the umbrella of demand-side management more generally, involves lighting, heating, and cooling. Efficiency efforts may also focus on energy-generating technologies, where the technology is improved to use less fuel inputs to generate energy outputs, or transmission technologies, where less power is lost through the electrical transmission and distribution process. Finally, efficiency is possible in the transport sector through refinements to the internal combustion engine or the introduction of other innovations such as electric vehicles.

Efficiency improvements offer several benefits. First, assuming that the cost of improving the efficiency of some good is less than the savings associated with reduced energy use, efficiency efforts save money, which can then be applied toward other goods, services, or development opportunities. Second, a reduction in energy consumption directly translates into a reduction of GHG emissions and other environmental pollutants, although the amount of decrease depends on the energy resource that powers the good. Third, improved electricity efficiency can reduce grid congestion and thereby save money on grid management and new energy construction.

The improvement of energy use or technological efficiency supports other energy-related goals as well. For example, if a country or individual uses less energy, assuming no rebound effect,[2] the result will be reduced GHG emissions and potentially enhanced energy security and less pressure to diversify its energy base.

A variety of energy efficiency programs are being deployed across the world, in both the public and private realms and at both national and local levels. These activities include the incorporation of energy efficient materials and products in standard building operations (e.g., light-emitting diode [LED] light bulbs, double-insulation window panes, or recycled stone, metals, industrial materials, and other

---

[2] A rebound effect is when someone uses more energy as a result of consuming a more efficient product. For example, if one buys a conventional hybrid vehicle with a much higher miles-per-gallon rating than his or her previous vehicles, then the amount spent on gasoline may decline. As a result of paying less to fill the tank, he or she may actually drive more miles than before.

"green building products") or methods such as smart-controls or targeted maintenance. These types of energy efficiency activities have the potential to reduce energy use and costs to businesses and residents for heating, cooling, and power. As a more concrete example, the rapidly growing city of Abu Dhabi is pushing the limit of water and electricity use as it continues to develop. The government understands that it must focus on conservation as it considers new production capacity and, as a result, is working to evaluate the country's demand for water and electricity and eliminate inefficient uses of these resources to save money and extend power and water supplies (RTI International 2013).

2.3.1.4 Energy Security

Energy security is defined as the ability to access reliable, affordable, and diverse energy (UNDP 2004), and it generally is conveyed in a strategic context at the national level. The IEA defines it as "the uninterrupted availability of energy sources at an affordable price" (IEA 2013b). The concept of energy security is far-reaching and includes aspects that are difficult to price or quantify such as reliability, affordability, and national safety. Energy security refers to geopolitical circumstances that do not compromise the integrity or safety of a nation, the quality delivery of energy services, and the use of diverse energy resources. Those countries that seek energy security aim to have a healthy energy mix and avoid overreliance on either any particular resource or nation.

It is difficult to predict the potential benefits associated with improving energy security, but these benefits are positive in that they give countries more autonomy and independence from other national governments, assuming that it does not come at the expense of reduced economic activity in global markets. Improving the reliability of national infrastructure or access to basic modern energy or reducing the vulnerability of a nation to foreign policy threats, for example, is indubitably important to all nations.

Denmark's bold goal of 100 % renewable energy by 2050 is an example of a country embarking on long-term plan for energy security. Denmark is making significant investments to shift its infrastructure and incentivize renewable energy industry development. The country is investing significantly, for example, in energy storage, distributed grid systems, and electric vehicles. Some countries are also pursuing energy security goals through natural gas extraction within their borders. For example, Russia stopped supplying natural gas to the European pipeline in the Ukraine in January 2009, one of the coldest times of the year in Europe. As a result, 20 countries in Europe had significant shortages of natural gas (MSNBC cited in Rao 2012). To respond to energy security issues such as this, Poland, for example, is developing regulations to promote the extraction of 70 years' worth of natural gas that could theoretically lower dependence on Russian gas (Moss 2012; Speak 2013).

2.3.1.5 Energy Poverty

Increasing access to energy is a means to reduce energy poverty. As discussed in Chap. 1, energy access is crucial for households and individuals and also for entire industries and businesses. Energy access allows school children more time in a day to study and write, members of a household time and resources to cook, and those within the medical profession the resources to provide quality medical care. Thus, energy poverty has deep implications for social development. Businesses, including industrial businesses, also need reliable, readily available, and affordable energy to run their operations. Some industries need to locate near affordable energy sources. For example, industries that run large computer server operations, such as Google, tend to locate these functions in temperate locations with readily available and affordable energy access (Bruns 2012). Electricity interruptions for these businesses are detrimental to their core services. Schools, households, hospitals, and manufacturers also need access to energy.

Energy access is so fundamentally important to basic human development and the development of nations that many countries have set national goals to achieve universal energy access. China and India, for example, both have universal energy access goals (Wang 2009; Ministry of Power, Government of India 2005). These goals can be reached through the provision of modern electricity access, either via the electrical grid or small-scale distributed applications that provide localized electric loads. Other ways to reduce energy poverty include cookstove programs that use energy efficient and cleaner sources of fuel for household cooking, rather than simple biomass, dung, crop residues, or charcoal. Woody biomass and these other resources tend to increase indoor air pollution, deplete local forests of trees, and consume a disproportionate share of women's time in the household for wood collection (Ezzati and Kammen 2002). Low-cost solar lanterns distributed to households and schools are another example of how low-emissions energy sources are being used to increase energy access.

2.3.2 Economic Development Goals

2.3.2.1 Increase Industry Growth

Regions, localities, and countries typically seek ways to expand their existing industry base and transition it into an emerging, globally competitive industry of the future. Diversifying a place's economic base reduces the overreliance on one or a few sectors and, in turn, reduces susceptibility to economic cycles or massive closures within those sectors. Industry growth is important to economic developers because the benefits from it spill over into other economic development goals. As industries grow, typically jobs and income to households and government jurisdictions also expand and generate more wealth for an area, which can fuel a positive cycle of growth and expansion. Industry growth goals in economic development are closely intertwined with goals to increase income and jobs, other economic development goals described later in this section. The rise of national economies in Southeast Asia is often cited by development economists as examples of countries that have launched themselves through industry growth into globally competitive markets and reaped the subsequent economic benefits (Amsden 2001; Wade 1990).

Industry growth efforts aim to reduce costs and improve information exchange and coordination between businesses within an industry and between business and government, respectively. The North Carolina Biotechnology Center, located in North Carolina, USA, is an organization established to lead an industry growth strategy. This initiative is supported by state government and houses an industry library, provides grants and loans to companies, and serves as a liaison office to government officials to strengthen understanding of the industry and its needs for growth by state leaders. Taiwan offers an example of an industry strategy that set out to reduce cost barriers for an emerging industry. Taiwanese officials sought to diversify its agriculture industry from sugar to orchid production. The government invested $65 million in supportive infrastructure such as a genetics laboratory, shipping and packing facilities, roads, quarantine facilities, water and electricity hook-ups, and low interest credit to farmers (Rodrik 2007, p. 104 citing Bradsher 2004, p. A1).

Efforts to boost R&D, innovation, and entrepreneurship in emerging sectors are also strategies that support industry growth. The city of Pittsburgh's Alpha Lab is a business incubator that fosters entrepreneurship and business growth in software applications. It coordinates entrepreneurs and investors in this sector and provides low-cost space and grants to promising start-up companies, and it aims to lay the foundation for a globally competitive software industry of the future.

2.3.2.2 Increase Income

Increasing income is closely related to industry growth. Investments to boost industry expansion include attracting foreign direct investment (FDI) and increasing exports to generate more income for a country or a locality. Strategies to increase GDP, FDI, or a local tax base are all examples of ways to increase a place's revenues. More income allows for greater investment in infrastructure, education and training, public health, and quality of life, all factors that propel further economic development. For example, Taiwan and South Korea became "export platforms" in the 1960s, as Wade (1990) describes as "importing capital and intermediate goods, adding further processing with cheap labor, then exporting" and have become "highly integrated economies, moving speedily into high-wage, high-technology activities" (p. 42).

Countries often use a suite of policies and regulations to increase income, ranging from export incentives, marketing and recruitment, import controls, tariffs, industry protection, foreign exchange and exchange rates to export processing zones. Somewhat related, but on a smaller scale, localities use different techniques to attract income to their region. Business and industry recruitment strategies are approaches used by officials to site a company outside of their region to their jurisdiction. Tax and finance schemes such as enterprise zones or tax increment financing are techniques local economic developers employ to revitalize or incentivize investments to targeted areas. Reduced taxes or favorable financing mechanisms are used to reduce overall costs to an individual company to attract several of them to colocate, thus generating an increase in overall income to a jurisdiction (Eisinger 1998). We discuss some of these financial mechanisms in more detail within the industry policy discussion in Chap. 4.

2.3.2.3 Increase Innovation and Entrepreneurship

Innovation and entrepreneurship are increasingly used as ways to transform economies from low-cost and low-wage to higher value-added types of production or distribution of products. Innovation rests on businesses pushing new technologies, products, or processes to markets. Innovation involves scientific or other discoveries that companies can use to expand a business line or start an entirely new line of business. Innovation can also lead to modifications of the business model itself. In recent years, this approach to innovation has expanded to a focus on economic development on innovation ecosystems, which includes interactions between actors and institutions as part of the innovation growth process.

In advanced economies, innovation tends to be closely linked to sophisticated technology development, but it has broader connotations in less developed economies. The World Bank stresses that innovation is not just linked to technology; it is something new that can readily be diffused to a region or country in which it appears that can create jobs and wealth or improve welfare (World Bank 2010, pp. 10–11 and p. 54). Innovation is "technologies or practices that are new to a given society ... not necessarily new in absolute terms" (2010, p. 4).

A mobile payment system based in Kenya, M-PESA, is an example of a successful innovation adopted at a large scale. This software application, which literally translates to "mobile money," allows people to transfer money by text message on their cell phones. This technology has revolutionized how Kenyans conduct their business and personal finances. According to *Bloomberg Businessweek* magazine, over 14 million Kenyans, or approximately 70 % of the country's adult population, have an M-PESA account (Greeley and Ombok 2011). M-PESA has increased access to finance, increased financial security, and enhanced benefits of business-to-business transactions by improving communication of market information and timeliness of payments between businesses. In addition, this service has allowed Kenyans to send or spend money for other purposes as well, such as for shopping, utility bills, taxi rides, or distant relatives (Graham 2010).

Innovation and entrepreneurship are linked: as new ideas are turned into innovations, these innovations need a business model and enterprise to help launch them into markets so that they can be widely disseminated. Somewhat distinct from innovation, entrepreneurship is also often viewed as a means for countries and localities to create their own employment opportunities and generate local income by "growing their own" market needs with local businesses and enterprises.

China's rural township and village enterprises (TVEs) offer a unique example of entrepreneurship strategies that played a key role in helping a country modernize. Throughout rural China, TVEs are small enterprises, either collectively owned by local residents or the government, that have significant local autonomy and strict budget constraints. These enterprises are credited as playing a significant role in Chinese economic development since the late 1970s, accounting for up to 19 % of average annual real growth rates from the late 1980s to 1990s; in 2000, estimates are that these enterprises accounted for as much as 47 % of total industrial output for that year (Fu and Balasubramanyam 2003, p. 27).

Another example of an entrepreneurship support organization is Techstars, which is one of the world's largest business accelerators. This program pools venture and angel investors to support entrepreneurs in their quest to take innovations to market. With an investment from Techstars of approximately $118,000 in start-up capital and business mentoring for each start-up, these companies then average $1.6 million in outside venture capital, raised after they exit the entrepreneurship support program (Techstars n.d.).

2.3.2.4 Job Creation

Job growth is often considered the ultimate endgame for economic developers. Employed workers that earn fair wages tend to make for more stable societies. Politicians, policymakers, advocates, and civic leaders of all backgrounds tend to seek ways to create jobs so that their populations benefit from greater stability. This goal is perhaps the most cross-cutting in that it is often the result when other goals are met. Job creation becomes a top priority among economic development goals in places with significant unemployment, experiences in sharp economic downturns, or mass lay-offs within a critical industry base.

Job creation strategies generally focus on immediate or medium-term ways to create employment opportunities so that individuals have the means to earn income gainfully to support themselves and their families. Job creation efforts are divided into two camps: those that generate employment and those that increase the employability of workers (Karnani 2009).

In times of economic crisis, governments often invest in large public infrastructure projects as a means to create immediate and large numbers of jobs to have a noticeable and significant impact. In the recent global recession, many countries used this tactic within their self-prescribed stimulus packages or applications to development banks. For example, in late 2008 Armenia requested funds from the World Bank to help the country weather the economic shocks from the global financial crisis. Sharp declines in GDP, FDI, and employment coupled with increases in poverty forced the country to apply for emergency funds to immediately begin a road improvement project of over 100 km. The resulting project was estimated to create 7,600 person-months of temporary jobs (Ishiria and Bennett 2010, pp. 1–2). Although not a long-term fix, these kinds of solutions help keep incomes and families afloat until the economy recovers. As a different example, during the recent recession in the United States many government-sponsored energy efficiency buildings programs were as much an effort to employ laid-off construction workers in rural and urban areas as an effort to increase energy efficiency.

A second type of job creation strategy is workforce training or mechanisms to enhance employability of the unemployed or underemployed. Education that helps workers retool their skillsets for jobs in newly competitive industries occurs in places where mid- or late-career workers are caught in an economic adjustment, where the techniques used are not in line with competitive industry practice. Youth employment is also an area of focus for job creation strategies. As of 2013, 290 million or a quarter of the world's youth—defined as 15- to 24-year-olds—are neither employed, in school, or in a formal training program (*The Economist* 2013). Efforts to address this epidemic typically focus on bridging the skills gap between what is taught in school and what business and industry need. Vocational and technical schools and industry-based training programs typically are used to help boost skills training in unmet demands. Examples range from South Korea's government-sponsored vocational "meister" schools for machine operators and plumbers to IBM's training school established in New York. A newly planned community outside of Ramallah in the West Bank aims to address this skills gap through virtual employment of its workers in its information technology hub, with international companies serving as a sort of "on-the-job training" for the well-educated but underskilled workforce there (Lawrence et al. 2009).

2.3.2.5 Reduce Poverty

Similar to job creation, poverty alleviation is often an underlying goal for most economic development strategies and of course closely tied to job creation and income generation goals. Efforts to boost incomes either for companies, governmental jurisdictions, households, or workers all ultimately aim to generate wealth. Greater wealth in the form of income and assets improves a locale's ability to invest in physical infrastructure (e.g., roads, ports, water, broadband Internet, and education and training) and a household's or individual's ability to pursue personal investments (e.g., healthcare, education, safe environment, and transportation). In regions marked by impoverished conditions, more resources to make infrastructural and personal investments help facilitate an environment that allows individuals to succeed and create economic opportunities and thus break the cycle of poverty.

A significant number of policies and programs aim to reduce poverty, many of which are deeply researched and highly debated in terms of their effectiveness. These policies and programs range from childhood malnutrition and maternal and child health programs, basic road construction, water and sewage infrastructure to microcredit programs that help families access affordable credit to generate income through entrepreneurial activities. Given that antipoverty efforts are so sweeping in nature, for the purposes of this book, we focus on efforts targeted more specifically at increasing income and wealth for households or individuals.

Employment and training programs for disadvantaged youth in urban slums are an example of a poverty reduction program. Funding support to help reduce the opportunity costs of young girls going to school and free or reduced-cost healthcare clinics are other examples of antipoverty programs.

2.4 Conclusion

EBED has several distinguishing features, the most prominent of which is its focus on advanced, efficient, and low-emissions energy. EBED is different from related disciplines in that it pursues joint energy and economic development goals, as outlined in the chapter above. It involves a more comprehensive and diverse set of actors, champions, and beneficiaries as a result of the more dispersed nature of the energy resources EBED supports. EBED also offers researchers and practitioners an integrated framework to align what are often considered disparate goals into a unified approach. Finally, EBED recognizes the role of governance, leadership, and stakeholder models in shaping its success.

EBED is a direct extension of the energy planning and economic development disciplines. As both founding disciplines have evolved through the years, a number of synergies in practice and objectives have emerged that mark the nexus of the EBED domain.

In the following chapter, we describe the process of designing, implementing, and evaluating an EBED program.

References

Amsden A (2001) The rise of the rest: challenges to the West from late-industrializing economies. Oxford University Press, New York
Barnes D, Floor WM (1996) Rural energy in developing countries: a challenge for economic development. Annu Rev Energy Environ 21:497–530
Bolton R (1991) "Place prosperity vs. people prosperity" revisited: an old issue with a new angle. Urban Stud 29:185–203
Boström M (2006) Regulatory credibility and authority through inclusiveness: standardization organizations in cases of eco-labelling. Organization 13:345–467
Bradsher K (2004) Once elusive, orchids flourish on Taiwanese production line. New York Times, August 24, p A1
Bradsher K (2010) China leading global race to make clean energy. New York Times, January 30
Bruns A (2012) Big data blitz. Site Selection Magazine, July. http://www.siteselection.com/issues/2012/jul/data-centers.cfm. Accessed 13 June 2013

References

C40 Cities, Climate Leadership Group (n.d.) A 12.9 km bus rapid transport system built in just 9 months at a cost of $2 million/km. http://www.c40cities.org/c40cities/jakarta/city_case_studies/a-129-km-bus-rapid-transport-system-built-in-just-9-months-at-a-cost-of-2-millionkm. Accessed 5 May 2013

Carley S, Lawrence S, Brown A, Nourafshan A, Benami E (2011) Energy-based economic development. Renew Sust Energy Rev 15:282–295

Carley S, Brown A, Lawrence S (2012) Economic development and energy: from fad to a sustainable discipline? Econ Dev Q 26:111–123

China Daily USA. Action to clean air. September 13, 2013. http://www.chinadaily.com.cn/cndy/2013-09/13/content_16966749.htm. Accessed November 8, 2013

Eisinger PK (1998) The rise of the entrepreneurial state: state and local economic development policy in the U.S. The University of Wisconsin Press, Madison

Energy for All (n.d.) http://www.energyforall.info/about/energy-for-all/. Accessed 24 April 2013

European Commission (2013) Climate action. http://ec.europa.eu/clima/policies/brief/eu/index_en.htm. Accessed 5 May 2013

Ezzati M, Kammen DM (2002) The health impacts of exposure to indoor air pollution from solid fuels in developing countries: knowledge, gaps, and data needs. Environ Health Perspect 110:1057–1068

Fitzgerald J (2010) Emerald cities: urban sustainability and economic development. Oxford University Press, New York

Frankfurt School—UNEP Collaborating Centre for Climate & Sustainable Energy Finance (2012) Global trends in renewable energy investment 2012. UNEP Collaborating Centre, Frankfurt School of Finance & Management, Frankfurt am Main

Fu X, Balasubramanyam VN (2003) Township and village enterprises in China. J Dev Stud 39:27–46

Gallagher KS, Holdren JP, Sagar AD (2006) Energy-technology innovation. Annu Rev Environ Resour 31:193–237

Graham F (2010) M-PESA: Kenya's mobile wallet revolution. BBC News Business, November 22. http://www.bbc.co.uk/news/business-11793290. Accessed 16 May 2013

Greeley B, Ombok E (2011) In Kenya, securing cash on a cell phone. Bloomberg Businessweek Magazine, September 8. http://www.businessweek.com/magazine/in-kenya-securing-cash-on-a-cell-phone-09082011.html. Accessed 16 May 2013

Hausmann R, Rodrik D, Velasco A (2008) Growth diagnostics. In: Serra N, Stiglitz J (eds) The Washington consensus reconsidered: towards a new global governance. Oxford University Press, Oxford, pp 324–355

International Energy Agency (IEA) (2013a) Energy policy of IEA countries—Germany—2013 Review. Executive Summary and Key Recommendations. http://www.iea.org/Textbase/npsum/germany2013SUM.pdf. Accessed 21 June 2013

International Energy Agency (IEA) (2013b) Energy security. http://www.iea.org/topics/energysecurity/. Accessed 6 May 2013

Ishihara S, Bennett CR (2010) Improving local roads and creating jobs through rapid response projects: lessons from Armenia Lifeline Roads Improvement Project. World Bank, Washington, DC. https://openknowledge.worldbank.org/handle/10986/11711. Accessed 15 May 2013

Kan H, Chen R, Tong S (2012) Ambient air pollution, climate change, and population health in China. Environ Int 42:10–19

Karnani A (2009) Reducing poverty through employment. University of Michigan, Ross School of Business Working Paper, Working paper No. 1132. http://deepblue.lib.umich.edu/bitstream/handle/2027.42/64053/1132_Karnani.pdf?sequence=1

King Abdullah University of Science and Technology (2013) About economic development. http://www.kaust.edu.sa/economic_development/about.html?submenuheader=0. Accessed 21 June 2013

Lawrence SE, Dajani A, Bowditch NH, Schwartz GM (2009) Leveraging knowledge assets in the first Palestinian planned community. In: IASP, The Research Triangle Park (eds) IASP World Conference on Science and Technology Parks: Future Knowledge Ecosystems—The Opportunity for Science and Technology Parks, Places and Partners, Raleigh, NC, June 1–4 2009

Malizia EE (1994) A redefinition of economic development. Econ Dev Rev 12:83–84

Ministry of Power, Government of India (2005) National electricity policy. The Gazette of India, February 12. http://powermin.nic.in/whats_new/national_electricity_policy.htm. Accessed 21 June 2013

Moss P (2012) Denmark's renewable energy goals wishful thinking? BBC News Science and Environment, April 8. http://www.bbc.co.uk/news/science-environment-17628146?print=true. Accessed 10 November 2012

Nissing C, Blottnitz H (2010) Renewable energy for sustainable urban development: redefining the concept of energisation. Energy Policy 38:2179–2187

Peters M, Fudge S, Jackson T (eds) (2010) Low carbon communities. Edward Elgar, Cheltenham

Rabe BG (2004) Statehouse and greenhouse: the emerging politics of American climate change policy. The Brookings Institution Press, Washington, DC

Rabe BG (2008) States on steroids: the intergovernmental odyssey of American climate policy. Rev Policy Res 25:105–128

Rao, V (2012) Shale gas: the promise and peril. RTI International Press, Research Triangle Park, NC

Rodrik D (2007) One economics, many recipes: globalization, institutions, and economic growth. Princeton University Press, Princeton

RTI International (2013) The Abu Dhabi diet: staying cool and saving green. http://www.rti.org/page.cfm?objectid=75D15272-5056-B100-3153F7F61192440D. Accessed 6 May 2013

Runci P (2005) Renewable energy policy in Germany. Pacific Northwest National Laboratory Technical Lab Report PNWD-3526. http://www.globalchange.umd.edu/energytrends/germany/. Accessed 23 June 2013

Späth P, Rohracher H (2010) "Energy regions": the transformative power of regional discourses on socio-technical futures. Res Policy 39:449–458

Speak C (2013) Poland rethinks its strategy for shale gas production. The Prague Post, June 12. http://www.praguepost.com/news/16488-poland-rethinks-its-strategy-for-shale-gas-production.html. Accessed 21 June 2013

Stimson R, Stough R, Roberts B (2006) Regional economic development: analysis and planning strategy. Springer, Berlin

Techstars (n.d.) http://www.techstars.com/. Accessed 16 May 2013

The Economist (2013) Generation jobless. The Economist, April 27, pp 49–52

Toman M, Jemelkova B (2003) Energy and economic development: an assessment of the state of knowledge. Discussion paper 03-13. Resources for the Future, Washington, DC

United Nations Development Programme (UNDP) (2004) World Energy Assessment Overview: 2004 Update. United Nations Development Programme, Bureau for Development Policy, New York

United Nations Development Programme (UNDP) (2005) Energizing the millenium development goals: a guide to energy's role in reducing poverty. United Nations Development Programme, New York

Wade R (1990) Governing the market. Princeton University Press, Princeton

Wang T (2009) Rural electrification in China: experience and lessons. Tyndall Centre Programme 4 Workshop. http://tyndall.ouce.ox.ac.uk/prog4/events/adaptationworkshop260409/tao-wang.pdf. Accessed 21 June 2013

Wong E (2013) Air pollution linked to 1.2 million premature deaths in China. New York Times, April 1. http://www.nytimes.com/2013/04/02/world/asia/air-pollution-linked-to-1-2-million-deaths-in-china.html?_r=0. Accessed 10 May 2013

References

World Bank (1996) Rural energy and development: improving energy supplies for two billion people. World Bank, Washington, DC

World Bank (2010) Innovation policy: a guide for developing countries. World Bank, Washington, DC

Zadek S (2011) Beyond climate finance: from accountability to productivity in addressing the climate challenge. Clim Policy 11:1058–1068

Chapter 3
Process and Approaches

Abstract Chapter 3 presents the EBED process of an EBED initiative and is intended to help guide policymakers and practitioners in designing and implementing EBED initiatives. The first three steps—stakeholder engagement; goal identification; and asset, needs, and gaps identification—are essential for ensuring the framework is used effectively. The remaining steps are selection of an approach, development of metrics, implementation, and monitoring and evaluation. Each step is described in detail and provides information for EBED stakeholders to tailor an approach so that it is most conducive to local needs, assets, and circumstances. This chapter concludes with a description of three fundamental factors that shape the design of an EBED project: the point of intervention, geographic scale, and the degree of transformation that the project requires relative to the status quo.

Like all development projects, EBED projects involve a change or evolution from one state or set of conditions to another. A well-defined process is therefore critical to navigate through these states of change. In the first part of this chapter, we present the process for EBED initiatives. Our discussion is intended to help guide practitioners and policymakers in designing and implementing EBED initiatives. Although each project is context and location specific, this process offers insights on elements that can be incorporated into all EBED initiatives.

The second part of the chapter addresses three different ways to approach EBED: the point of intervention, the geographic scope, and the scale of transformation. These three factors distinguish the differences in EBED approaches across initiatives. Our intent in highlighting these different dimensions is to help those leading EBED initiatives be more intentional in their approach and to enhance the program's ability to meet the goals and objectives.

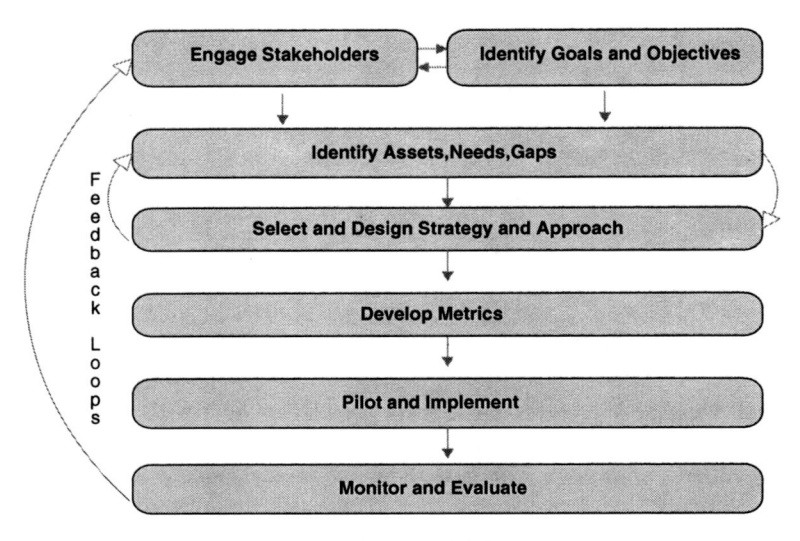

Fig. 3.1 EBED process (adapted from Carley et al. 2012)

3.1 Process

The EBED process consists of seven elements, as diagrammed in Fig. 3.1:

1. stakeholder engagement
2. goals and objectives identification
3. assets, needs, and gaps analysis
4. strategy and approach design
5. metric development
6. pilot and implementation
7. monitoring and evaluation.

Similar to many economic and community development projects, the EBED process typically begins with *engaging stakeholders* who have identified unmet needs within their community or country. Through this engagement and exchange of ideas, EBED practitioners *identify goals and objectives* for the EBED program. Sometimes, however, this process happens in reverse order: selected champions of an EBED initiative may define goals and objectives at the outset and then engage a broader set of stakeholders to advance the program or project.

Once goals and objectives are established, practitioners identify and analyze *assets, needs, and gaps* within the target area. Documentation and analysis of the assets, needs, and gaps provide a deeper understanding of the conditions, opportunities, and obstacles that EBED stakeholders and practitioners face. The next step is to *select and design the strategy and approach* so that it is tailored to fit the circumstances of the region. These last two steps are often an iterative process, whereby the design of the strategy uncovers or necessitates a revision of the assets, needs, and gaps, which then may require adjustments in the design phase. The feedback loop between these two steps is demonstrated with smaller arrows in Fig. 3.1. Once program designers establish these elements, *metrics are identified* and baseline conditions are measured. In the next step, a *project is launched into a pilot or full-scale implementation*. Once the EBED project is underway, practitioners and analysts *monitor and evaluate* the initiative to track its progress. These steps are described in greater detail below.

This entire process is also part of a broader feedback loop, where an understanding of the project process and outcomes informs stakeholder participation or project goals and objectives, which leads to modifications in strategy and so forth. It is also important to note that, at any stage, EBED practitioners can return to previous steps to revisit and revise their process.

This process varies in composition across different EBED initiatives, where elements of one step may be more relevant for a particular initiative than others. For example, a social entrepreneur may not need to engage a broad group of stakeholders nor conduct a comprehensive workforce or innovation assessment. She may, however, need to engage relevant suppliers, government, or utility officials and have an understanding of workforce capabilities to launch her enterprise. She will also need to monitor outputs, sales, and revenues to assess the enterprise's viability over time. Therefore, she follows the basic steps and simply tailors them to fit the needs of her endeavor.

As another example, those involved in an effort to propel an emerging renewable energy industry, such as solar or wind, may undertake a different series of tasks as part of their EBED process. These practitioners will first consider obstacles and opportunities and discuss these conditions with industry and innovation leaders to define appropriate goals and objectives. These goals and objectives will then inform the design of initiatives and corresponding policies to support the industry's growth. Obtaining data on the availability of the resource, market potential, generating facilities, and workforce will also help shape the EBED initiative and supporting policies. Once the initiative is underway, frequent evaluations can help practitioners and analysts decipher what is working within the EBED design and what is not and can thus inform program refinement.

We believe that it is important to consider each step of the process because EBED efforts may fail if important steps are overlooked. An effort to install wind or solar farms in a region that does not engage relevant stakeholders can stall if communities around the proposed sites are unaware of how their local landscape will change or other economic impacts that the proposed project may have, such as changes in energy costs, industry development, or job creation. Similarly, the same region may be set to develop wind or solar industries but has not yet collected and analyzed resource endowment and industrial composition data. Thus, the region may misjudge its location's assets and economic competitiveness position.

The first three steps of the process are important because they establish the intent of the EBED effort from which EBED designers and stakeholders can construct an approach. The last four steps follow according to the intentions established in the first part of the process. It is thus important to be clear and purposeful when considering stakeholder engagement; goals and objectives; and identification of assets, needs, and gaps.

3.1.1 Engage Stakeholders

Stakeholder engagement takes many forms in terms of who, how, and with what frequency groups of concerned entities participate in a planning process. Nevertheless, stakeholder engagement is fundamental to creating programs and policies that are grounded in the needs of the target region or population. With an EBED project that involves local implementation, it is difficult to understand the context without engaging stakeholders who are critical to the process in that area. Engagement with stakeholders therefore helps provide EBED practitioners with a stronger understanding of the cultural, political, and social barriers that may be unforeseen to someone who does not live there. Stakeholder engagement is also important because it builds champions within the community, which is often a critical factor for program success, particularly one that affects many individuals or organizations. Champions help the program gain traction and achieve sustainability over time. Engaging a variety of stakeholders also has the potential to elicit viewpoints across many individuals and organizations and therefore provide multiple perspectives on the EBED initiative.

In an EBED setting, key stakeholders could include government officials (e.g., energy planners, economic policy advisors, environmental experts), utility representatives, industry representatives, technology developers, entrepreneurs, international donors, educators, advocates, nonprofits, civic leaders, representatives from the target population, and laborers. This list is not intended to be exhaustive, merely representative of a range of stakeholders.

This step of the EBED process sometimes occurs in parallel with the identification of goals and objectives, particularly when a subset of individuals identifies goals and objectives for a project and subsequently takes the ideas to a community of stakeholders to gain a broader base of support for the project. Without stakeholder engagement, projects can run afoul because they catch individuals with vested interests in the outcome of the project off guard and with inadequate information to make sound decisions about if and how to advance the project.

3.1.2 Identify Goals and Objectives

In this step of the EBED process, practitioners, policymakers, or stakeholders clarify the goals and objectives of the initiative. As discussed at length in Chap. 2, the EBED goals are provided here for reference:

Energy policy and planning	Economic development
• Diversify energy sources	• Drive industry growth
• Reduce greenhouse gas emissions and related environmental impacts	• Increase innovation and entrepreneurship
• Increase energy efficiency	• Increase regional income
• Increase energy security	• Decrease poverty
• Reduce energy poverty	• Create jobs

We distinguish between goals and objectives: goals are broader targets and objectives are specific intentions for each project. For example, an EBED energy service provision initiative may set guiding goals of a reduction in energy poverty and an increase in income. The more specific objectives might be to increase electrification in a certain number of households, establish new small businesses that sell energy systems and replacement components, coordinate credit or loan arrangements with a local bank, and work with the government to reduce regulatory or administrative hurdles that could delay the initiative.

EBED initiatives blend multiple goals from its energy and economic development foundations, involving at least one goal from each discipline. This step is important because it enforces alignment between goals, objectives, and metrics, which helps the program designer shape initiatives based on the underlying purpose of the effort.

3.1.3 Identifying Assets, Needs, and Gaps

Data collection and assessment of EBED assets, needs, and gaps are critical for an EBED initiative. Data and data analyses help practitioners and other decision makers ascertain how the assets of a region best align with the program goals and objectives. Data also inform EBED practitioners about the gaps in their EBED asset base so that practitioners can directly address shortcomings and gaps. Figure 3.2 displays illustrative data points that an analyst might gather to better understand the conditions within a targeted community or region. We divide the data categories into business and industry, market growth potential, energy resources, sociodemographic circumstances, and environmental conditions. These data can be collected from existing resources such as energy, industry, or labor offices that already collect this information or, if needed, through surveys, interviews, or any other primary data-gathering effort. Quantitative data can also be supplemented with qualitative interviews to help practitioners and analysts gain a deeper understanding of issues for which quantitative data provide an incomplete picture. Further research to expand and refine appropriate data sets will depend on the specific goals and objectives identified in the prior step and on data availability.

Business and Industry	Market Potential	Energy Resources	Household/ Sociodemographic	Environment
• Number of firms • Firm concentration • Firm growth • Business creation/loss	• Growth sectors • R&D activity • Patent filings • Absorptive capacity • Entrepreneurial capacity • Access to capital	• Natural resource endowment (e.g., wind, solar, natural gas, waste, bio-materials, hydro) • Other energy assets (e.g. nuclear, electric vehicle charging infrastructure) • Energy usage rates • Utility capacity	• Income • Poverty rates • Employment/ unemployment rates • Electricity availability • Energy access rates	• Greenhouse gas emissions • Forest, water, air, or agricultural degradation • Pollution rates • Sea level rise

Identify Assets, Needs, Gaps

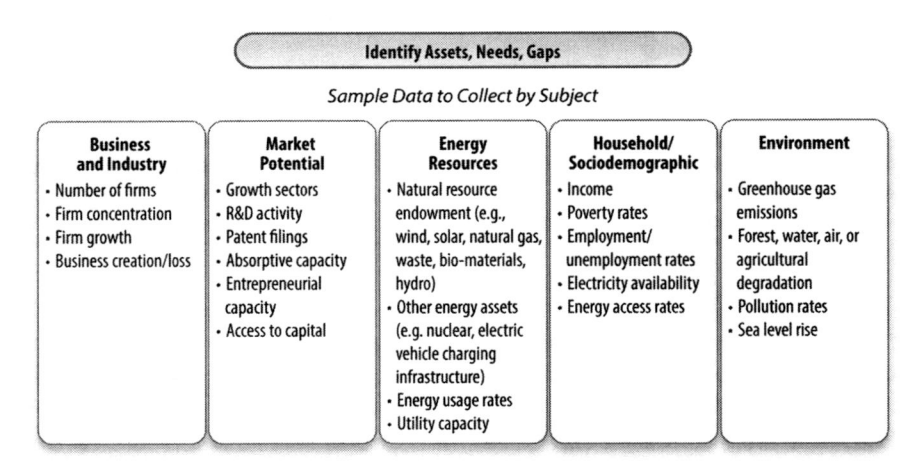

Sample Data to Collect by Subject

Fig. 3.2 Illustrative data points to help assess assets, needs, and gaps

After data collection and analysis, it is important to assess what the data reveal and the degree to which either the EBED idea aligns with the region's or market's ability to produce and absorb it or needs within the region can feasibly be met with an EBED intervention. Regardless, if one is starting with an idea or with a need, understanding how the asset base aligns is a critical step for project leaders or stakeholder groups to undertake. Data analysis also provides insights on which existing resources or organizations may help a project succeed or which barriers may prevent project success. Along these lines, determining whether an intervention has close alignment, moderate alignment, or limited alignment to current or projected conditions helps stakeholders decide how to best advance EBED goals and objectives. This type of exercise also helps practitioners and policymakers understand the level of effort required for a location to achieve its goals. The three degrees of alignment are as follows.

Close alignment. An EBED effort can be classified as "close alignment" when it relies on an asset base that is closely aligned with target needs. For example, regional economic development officials may focus on expanding a supply chain in the solar industry. The region of focus has several manufacturers of solar panels, demonstrated research and development activities in area universities and within the private sector, and demonstrated market success in deploying the solar-related products to regional or global markets. Another example is a city establishing a plug-in electric vehicle infrastructure to support the demand for electric vehicles by local residents and businesses. Finally, a rural region with building stocks ripe for energy-efficiency improvements and an underemployed but trained construction workforce also demonstrates close alignment for an energy-efficiency retrofit EBED project.

Moderate alignment. An EBED effort can be classified as "moderately aligned" when the target area has existing assets, but they need to be reoriented to meet current or projected market demands. For example, in a moderately aligned region, businesses may emerge with products and services that have potential for increased market penetration, but appropriate financing mechanisms to scale the production of associated products are scarce. This lack of finance may be coupled with a lack of understanding about how to export these products to markets that demonstrate potential for demand. With increased attention and coordination efforts to connect this company to potential funders and existing services for export promotion (e.g., training or expositions), this company could be well positioned to scale its low-emissions product to market. Another example is a region with some innovation, businesses, and utility-related assets that intends to shift to greater deployment of smart grids, but the government or utilities are not motivated enough to shift from business as usual to a smart grid infrastructure. With either greater convincing or a change in leadership, this alignment can become "close" and thus pave the way for the more likely adoption of an EBED initiative. In this example, the assets to deploy the smart grid technologies into electricity markets exist; institutional support is just lacking. With time, effort, and encouragement, these assets have the potential to align.

Limited alignment. Approaches in this category seek to advance a set of EBED goals but have few natural resource, workforce, or business-related assets from which to build an emerging economic base of EBED activity. Efforts that fall within this description are often subject to the tendency to "run in packs" (Fitzgerald 2010, p. x) where regions or states chase the most recent endeavor that causes excitement among practitioners, politicians, economists, and energy analysts. Approaches with limited alignment may aim to build off of what does exist to encourage market entry or to scale up those activities that demonstrate a potential for success in EBED sectors. It may also be helpful for EBED stakeholders in these circumstances to assess if an opportunity exists to create a stronger policy environment that may be more likely to foster EBED approaches. Places with limited alignment may need to consider either longer time horizons for EBED initiatives to take hold and/or greater investments upfront to strengthen the EBED asset base for future development.

3.1.4 Select and Design Strategy and Approach

Once goals, objectives, assets, needs, and gaps have been identified, the next step of the EBED process is to design a strategy or approach. Figure 3.3 offers examples of EBED projects and initiatives. The examples were chosen to illustrate the range of EBED initiatives that exist across the globe and represent a range of EBED goals, assets, needs, and gaps. These examples cover workforce development, clean energy adoption, technology commercialization and entrepreneurship, energy diversification, and exports. Additional and far more detailed examples are discussed in the second part of the book in Chaps. 7 through 9, which focus on specific case studies.

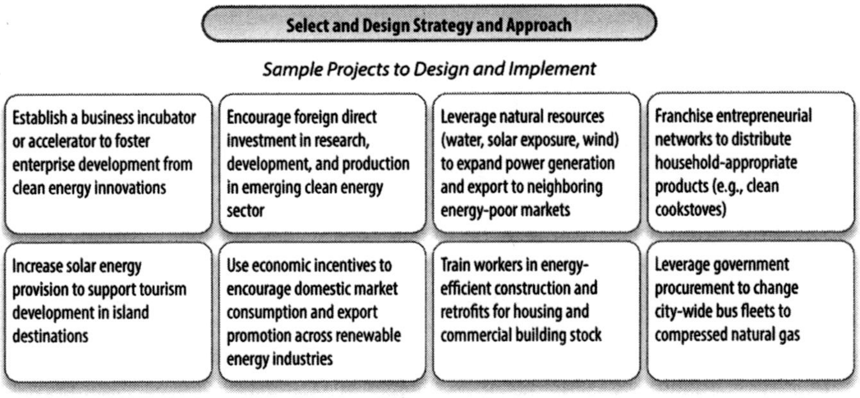

Select and Design Strategy and Approach

Sample Projects to Design and Implement

Establish a business incubator or accelerator to foster enterprise development from clean energy innovations	Encourage foreign direct investment in research, development, and production in emerging clean energy sector	Leverage natural resources (water, solar exposure, wind) to expand power generation and export to neighboring energy-poor markets	Franchise entrepreneurial networks to distribute household-appropriate products (e.g., clean cookstoves)
Increase solar energy provision to support tourism development in island destinations	Use economic incentives to encourage domestic market consumption and export promotion across renewable energy industries	Train workers in energy-efficient construction and retrofits for housing and commercial building stock	Leverage government procurement to change city-wide bus fleets to compressed natural gas

Fig. 3.3 Example EBED projects

3.1.5 Identify Metrics

The next step in the EBED process is to select metrics, based on the design and goals of the EBED project, to establish a baseline of conditions that currently exist and to track progress over time. Tracking progress enables practitioners and investors to assess if target impacts from the project goals are being achieved and, if not, to identify why. Some EBED program designers and funders may view this step as an afterthought to the process, instead of as a step linked to project design. We view this step as a critical part of the process and one that EBED program designers should plan for early in project conception and planning. Establishing outcome metrics, a research design, and an evaluation plan provides essential information to the program stakeholders about desired program outcomes, progress made, and possibilities for improvement. Moreover, data on EBED program metrics generate information for the wider EBED community on what works and what does not in different contexts and why. This information, in turn, can help other EBED practitioners tailor different EBED programs to respective markets and communities. Appropriate metrics depend on the type of program, targeted populations or markets, and goals that are selected as important for the program. Because of the importance of evaluation to EBED, we devote Chap. 5 to this topic.

3.1.6 Pilot and Implement

Once decision makers design the initiative and establish target metrics, they can then launch the initiative as either a pilot effort or full-scale program. The choice depends on factors such as funding, project readiness, geographic scale, and program scale. Pilots are often useful for social entrepreneurs, who can benefit from test cases that reveal what works and what does not, as they bring their products or services to market. This approach helps save resources in the event that a pilot program reveals glitches. For example, as discussed more extensively in Chap. 7, Nuru Energy is a company that produces and distributes solar lanterns. Nuru Energy first piloted their solar lantern schemes in select villages in rural Rwanda before they scaled it across other Sub-Saharan African countries and in India. In contrast, an effort as sweeping as the ARRA of 2009 injected stimulus funding into the U.S. economy during the onset of the recent recession. Thus, the government had no time to pilot such an effort; its intent was to address economic and energy needs across the United States immediately. The leadership of each project will have to decide if pilots are appropriate for meeting project goals. The larger and more important point of this step is that it is the part of the process that transitions the project from the conceptual stage into an active project.

3.1.7 Monitor and Evaluate

One of the most critical steps in the process is to monitor and evaluate the project's progress. With metrics in place and a baseline established, incorporating methods to collect and analyze data over time helps those directing and leading the effort identify the aspects of the project that are working and those that are not. Data gathered through evaluations also provide information to EBED practitioners and stakeholders about how to improve other steps of the process. Thus, quality communication, transparent and accurate data, and—when program budgets allow—independent evaluations are critical to this step in dispersing quality information back into the EBED process.

3.2 EBED Approaches

We devote the next half of this chapter to examining three fundamental factors that shape the design of an EBED project. The first factor is the point of intervention or the unit that the EBED initiative targets. The second factor is the geographic scale of the project. Is the initiative designed for a country, district, state, city, village, or neighborhood? The third factor is the degree of transformation that the project requires, relative to the status quo.

3.2.1 Point of Intervention

The point of intervention refers to the general target that the EBED initiative is designed to affect. It is important because each point suggests a different

Fig. 3.4 Points of EBED intervention

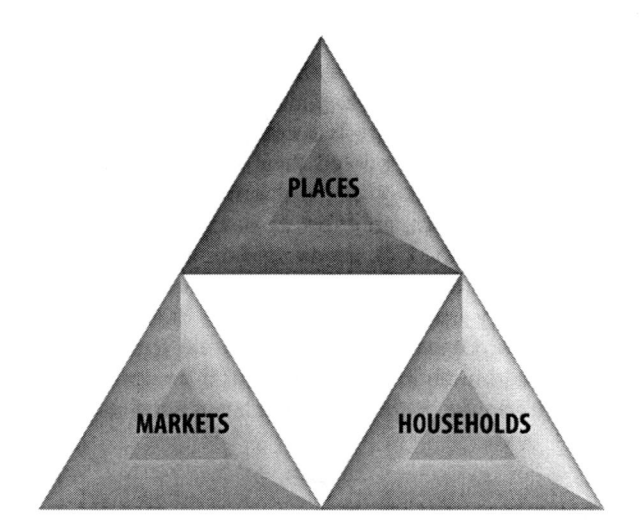

possibility for scaling the EBED effort and for determining which goals and objectives can be achieved. The majority of EBED initiatives fall within one or more of the following three categories: place based, market based, and household based. These three points of intervention are displayed in Fig. 3.4.

These categories are designed to help organize, develop, and frame the discussion on EBED approaches. These three classifications are not mutually exclusive; elements from one type of approach can cross over into another category. A project with more than one kind of approach would self-identify on this triangular figure not on a corner but somewhere on a line between two units or in the middle of the triangle if it contains elements of all three. A project such as the invention and dissemination of low-carbon cookstoves to poor households, for example, may lie between market and household because the project uses market-based approaches to target improvements for households. Other EBED programs may be even more comprehensive. The national SARi aims to generate local demand for renewable energy (i.e., target individuals and households within the country), expand its products to global markets for industry growth (i.e., a market-based approach), and target all activities within the program to benefit South Africa at large (i.e., a place-based approach). Other efforts may be more simplified. A program to introduce energy efficiency retrofits into building envelopes in urban neighborhoods is primarily a place-based approach that may also affect households. Therefore, this example is a place–household intervention. We provide more detail about each point of intervention and also relate this categorization concept to the case study descriptions in Chaps. 7 through 9.

3.2.1.1 Place-Based Approaches

Place-based approaches target a specific location, either a country, city, town, or other kind of designated district. These approaches tend to be designed and implemented by individuals who either work on behalf of a location (e.g., publically elected officials, government appointees, or regional economic developers) or individuals who have a vested interest or passion to enhance the economic or energy-related performance of a place (e.g., international donors, nonprofit organizations, citizens, residents, and other civically engaged stakeholders).

An EBED approach that is place-focused tends to work toward location-specific objectives to satisfy energy needs or to develop industry or entrepreneurship-focused growth strategies. Energy needs may include, for example, the provision of basic electricity to those in need or the development of distributed generation projects for a specific neighborhood or village. Energy needs may also include targeted energy efficiency projects in a specific location. Place-based growth strategies also involve efforts that aim to increase income, industry growth, or entrepreneurship in a specific region. These efforts may require the expansion of already existing business or industries; the fostering of business start-ups with potential to provide employment and growth into the future; or enhanced coordination between private, public, and education sectors.

Place-based growth strategies are based on the assets and market demand within a specific locale. These kinds of approaches require EBED stakeholders to identify their objectives and goals and then assess the degree to which these objectives align with markets, assets, services, and demand within their location.

An example of such efforts is an energy-specific industry cluster strategy, such as the smart grid cluster in the Research Triangle Region of North Carolina, in the United States. In this example, the region's private sector has coordinated with state government and area colleges and universities since 2012 to develop and expand a niche industry sector around smart grid activities and to market the region as a leader in this field. Another example from the same state is an effort to enhance energy efficiency in homes in the rural eastern region of the state. These efforts target under- and unemployed construction workers and hire them to conduct retrofits in poor households. Ultimately, the program aims to generate job opportunities and reduce energy costs for residents in these rural areas.

3.2.1.2 Market-Focused Approaches

Market-based approaches tend to focus on developing products, technologies, innovations, and techniques that can be absorbed readily into new markets to facilitate the growth of market share over time. Market-based efforts aim to enhance energy efficiency, expand energy provision, and provide cleaner alternatives for energy consumption. Regardless of the sector in which they work (i.e., government, nonprofit, higher education, or business), entrepreneurs, business leaders, technology developers, and innovators tend to drive this type of approach with support from policymakers, donors, and other stakeholders.

Innovation, product and process adoption, entrepreneurship, and trade are the central traits of market-focused approaches. Innovators create new technologies, products, or ways of doing business that, when introduced to relevant markets, change how and when energy is used in ways that achieve EBED goals.

The global demand for energy technologies that are advanced, efficient, and low-emissions is robust and growing. Markets for these emerging energy technologies offer significant opportunities for countries that invest to reap profitable returns. China, for example, is seizing such opportunities with dedicated investments in selected energy resource and technology markets because they have significant energy needs, markets ready to absorb these innovations, and the educated workforce and institutions to produce and disseminate these products. Many emerging economies are vying to be a "first mover" in energy markets and set themselves up to be at the global forefront in competitiveness for energy innovation. For example, investments by China in solar panel development have put the country at the forefront of global competitiveness in solar panel production. National industrial policy that uses subsidies, R&D, and manufacturing support have helped enable this first mover status (Bradsher 2009; Bradsher 2011; Vijayenthiran 2010).

Market-based approaches offer other potential benefits besides enhanced global economic competitiveness. Market development of energy products and services also supports industry growth strategies or other local economic development initiatives. Market-based approaches can also be applied to help a region achieve its energy goals, such as an increase in diversification of energy sources or enhanced energy efficiency.

In Chap. 4, we discuss in more detail the technology, innovation, and entrepreneurship policies that particularly are supportive of market-based approaches.

3.2.1.3 Household-Focused Approaches

The third type of EBED approach is tailored to individuals at the household level. This type of approach typically aims to reduce poverty, increase household income, increase energy provision, reduce energy poverty, increase cleaner sources of energy, and reduce household costs for energy through weatherization and energy efficiency applications. Implementers of these kinds of approaches are typically social entrepreneurs, nonprofit organizations, foundations, international donors, advocates, and government entities.

Household-based EBED approaches target individual needs and opportunities, such as job creation, workforce training, or skills upgrades for individuals seeking to climb the employment ladder from low-skill, low-wage jobs to employment opportunities that require more advanced skills and offer higher pay. Other initiatives that assume a household approach may simply aim to reduce the negative impacts of poverty. For instance, a waste-to-energy project that transforms methane from a large urban landfill into energy may employ 10–20 people from the neighborhood for a decade. This type of project targets employment in a high-poverty area, where job opportunities are slim, and provides an opportunity to increase household income, provide jobs, and build an employment track record for individuals in neighborhoods with limited employment options.

Household weatherization programs offer another example of this kind of EBED approach. The ARRA included skills training in weatherization techniques for laid-off construction workers. Many households targeted for upgrades through this program were low- to moderate-income earners who would eventually reap cost savings from cheaper energy bills. In developing countries, extensive efforts to replace cookstoves and diesel fuel used in homes also represent EBED programs that target households. These efforts aim to provide a cheaper and cleaner way to cook for the family, which reduces costs, but just as important reduces the health risks associated with dirty air to the mother and children who work within the home most of the day. Many social entrepreneurship efforts to create small enterprises and franchises that disseminate solar-powered lighting to remote rural villages in developing countries also embody household- or person-focused EBED approaches.

Each point of intervention—household, market, or place based—is relevant to the discussion of the EBED goals outlined in Chap. 2. Figure 3.5 demonstrates the potential alignment of all five goals with the three points of intervention. The three points of intervention are represented by the triangles at the top of the diagram. Goals are situated underneath them, where the green bars indicate economic development-oriented goals and the blue ones indicate energy policy and planning goals.

3.2.2 Geographic Scale

The second factor of EBED approaches is the geographic level at which an initiative is applied. The geographic level of intervention is an important consideration because EBED programs at more macro or national levels can lose sight of relevant activities at the local level. In the same vein, some EBED practitioners who work on local projects do so in isolation from national-level policymaking or other activities.

As the case studies in Chaps. 7 through 9 illustrate, EBED approaches are employed at both the national and subnational levels. National policies are important to support and promote more comprehensive EBED approaches. Incentives to induce certain behaviors and market exchanges directed by this level of government can have significant influence on how households and businesses make energy- and economic-related choices. Policy options, such as how to supply energy resources and encourage industry or employment growth, can affect the energy and business climate that influences which, and to what degree, types of energy technologies or industries flourish.

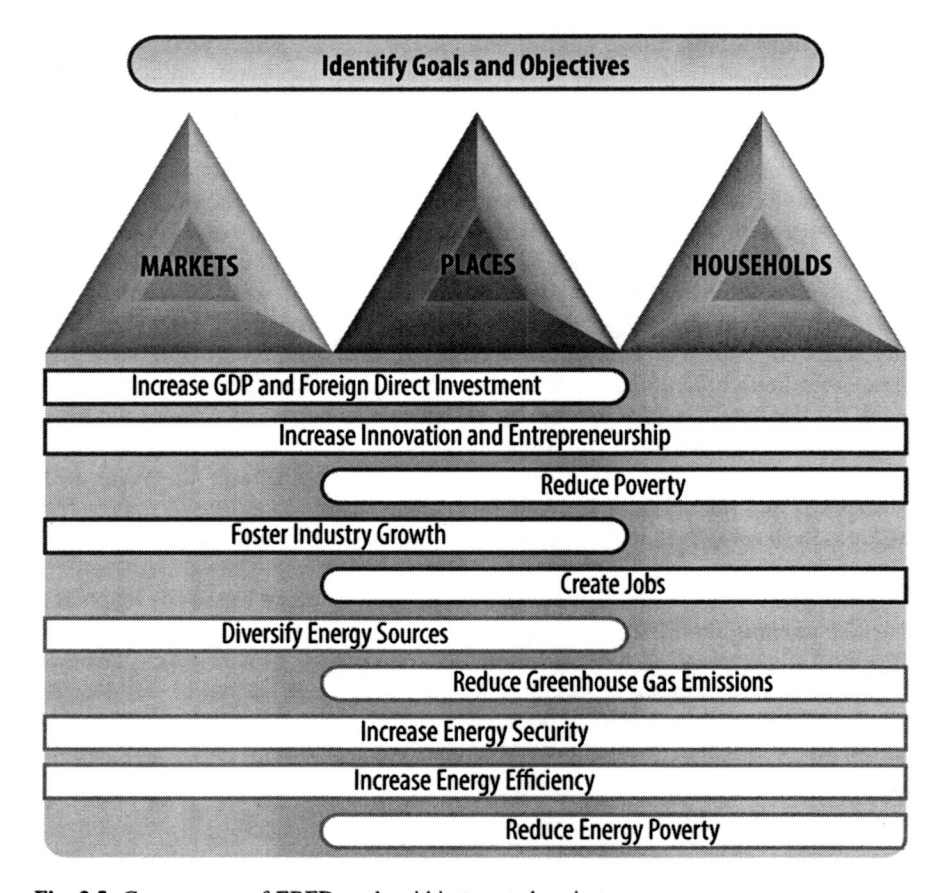

Fig. 3.5 Convergence of EBED goals within targeted projects

In contrast, localities play a key role in EBED because, in some countries, actors and institutions can organize projects and policies faster than national governments. For example, in the United States, states and localities are often referred to as "laboratories" for testing policies or program ideas. If these ideas become proven, the federal government may adopt elements from successful subnational programs and scale them to meet a country or regional context. In this role of subnational governments serving as laboratories for testing new ideas, these subnational EBED practitioners have a critical responsibility in demonstrating EBED success and failure so that replication can be easier with good documentation of what worked and what did not. Local approaches are also important when an initiative must be tailored to local circumstances and where a "one size fits all" approach to energy and development will not work well.

Most countries have some level of EBED activity that occurs at both the national and subnational levels. These programs either function as distinct efforts with little overlap or coordination between them or they operate in a compatible and complementary fashion. In some instances, national and subnational approaches can create complications. When both kinds of geographic approaches are in place, a lack of coordination between the approaches can result in confusion, inefficiencies, and perhaps redundancy. For instance, a social impact investor could be in the process of successfully distributing solar-powered lanterns for household lighting when a large international donor creates a widespread program to do the same. Without effective coordination and communication, these programs could undermine each other's success at worst and limit the potential to maximize positive impacts at best.

In places with subnational programs but no national strategies, local efforts can be difficult to scale because of a lack of wider awareness and support about energy-based economic development. For example, it would be difficult to scale the renewable energy and natural gas industries without the support of national-level policies that encourage the growth in these markets. However, national policies not tied to districts, states, cities, and communities will be difficult to implement effectively over time without the engagement and participation of subnational actors and institutions. For instance, if a national developer constructs a large wind farm without engaging the local workforce or economic developers, the region could miss EBED opportunities. Workforce and economic development specialists can help maximize economic potential from the associated job growth and supply chain needs resulting from the wind development. Capitalizing on these potential synergies between national and subnational networks is important to the success of EBED efforts.

3.2.3 Scale of Transformation

The final factor that characterizes EBED approaches is the transformation scale, or the degree to which an EBED initiative disrupts or changes energy and economic development behaviors or markets. For this concept, we borrow from Joan Fitzgerald, in her book *Emerald Cities* (2010), in which she defines the nature of change that a development approach may require (pp. 14–15). Fitzgerald identifies three types of transformation: linking in nature, transformative, or leap-frog past existing ways.

As the term implies, linking strategies generally connect entities and markets that are already present to create expanded opportunities. Linking strategies involve EBED practitioners identifying where latent synergies lie and working to build stronger connections among existing approaches and strategies after they assess existing assets. Two examples of linking strategies provided by Fitzgerald and apt for the discussion of EBED are efficiency retrofit programs for residents in need and a state industry strategy based on an already burgeoning clean energy sector.

Fig. 3.6 Scale of transformation

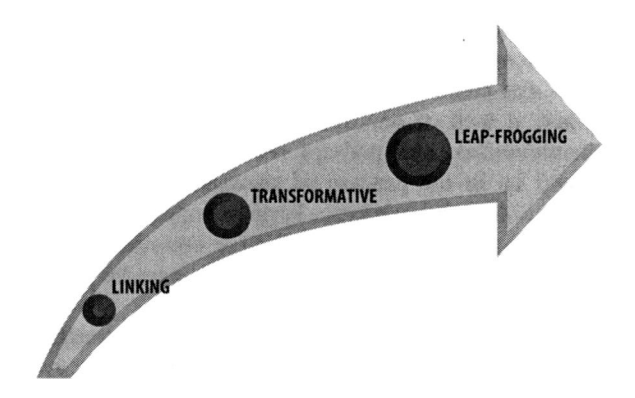

The next level of the transformation scale is transformative strategies. These kinds of strategies help businesses or entrepreneurs expand into new energy markets and modify existing practices to respond to emerging market conditions, economic opportunities, and the potential for industry growth. An example of a transformative strategy is a firm that produces a standard household appliance, such as an air conditioning unit, which it then alters to increase the efficiency and reduce the harmful emissions that are emitted through the production and use of the product.

When economic development strategies create entirely new market sectors, they "leap frog" over the existing products or services. Leap-frog strategies are the boldest and often riskiest type of approach trajectory because they involve the development and deployment of technologies that are new, untested, and thus more unpredictable in their success. China's development of an enhanced supply chain for the solar sector is a prime example of a leap-frog approach. On a smaller scale, a social entrepreneur may introduce bicycle-powered lighting in rural African villages that currently have no electricity.

We summarize this degree of transformation concept in the image depicted in Fig. 3.6, where an intervention can range from linking in nature on the bottom left to leap-frogging on the upper right side of the image, with transformative situated in the middle. We stress that there is not a better or worse place to be on this scale. More simply, a consideration of this factor helps the EBED program designer and implementers plan for the level of effort and risk they are willing to undertake.

3.3 Conclusion

Although the EBED process is similar to many others, the deliberate way in which it integrates energy and economic goals within a coordinated framework is important. This process enforces outcomes from an EBED process to be mutually beneficial to energy development and economic development. The first three steps—stakeholder engagement; goal identification; and asset, needs, and gaps identification—are essential for ensuring the framework is used effectively. The remaining steps provide valuable information for EBED practitioners to design and refine approaches that will work in their unique set of circumstances so that the EBED initiative is most conducive to local needs and circumstances. In the next chapter, we describe policies that enable, accelerate, and regulate the activities supportive of EBED.

References

Bradsher K (2009) China vies to be world's leader in electric cars. New York Times, 1 Apr 2009
Bradsher K (2011) Trade war in solar takes shape. New York Times, 9 Nov 2011
Carley S, Brown A, Lawrence S (2012) Economic development and energy: from fad to a sustainable discipline? Econ Dev Q 26:111–123
Fitzgerald J (2010) Emerald cities: urban sustainability and economic development. Oxford University Press, New York
Vijayenthiran V (2010) Chinese government pledges $15 billion for support of "green" cars. Green Cars Report, 10 Aug 2010. http://www.greencarreports.com/news/1048137_chinese-government-pledges-15-billion-for-support-of-green-cars . Accessed 30 April 2013

Chapter 4
Supportive Policies for Energy-Based Economic Development

Abstract EBED efforts are often implemented in conjunction with supportive policies and a conducive policy environment. Policies discussed in this chapter either enable and accelerate EBED activities, reduce the negative externalities associated with energy development, or strike some combination thereof. This chapter presents seven categories of policies that have the potential to advance EBED initiatives: technology innovation policies, technology adoption and commercialization policies, entrepreneurship policies, industrial growth policies, workforce development policies, climate and environment policies, and planning policies. Each of these policy categories is associated with a set of specific tools and techniques, as described in this chapter.

EBED is often accompanied by policies and a supportive political environment. Policies are used to enable, accelerate, or regulate desired action, behavior, and outcomes. This chapter provides an overview of public policies that have the potential to advance EBED initiatives. We identify seven common categories of EBED policies:

- technology innovation policies
- technology adoption and commercialization policies
- entrepreneurship policies
- industrial growth policies

- workforce development policies
- climate and environment policies
- planning.[1]

Each of these policy categories is associated with a suite of specific tools and techniques, which results in a menu of policy options for different locales to consider when designing and implementing their EBED activities.

All of the policies identified under these seven categories aim to enable and accelerate EBED activities or to reduce the negative externalities associated with energy development, or some combination thereof. All of these policies support the common EBED goals outlined in Chap. 2. They also guide program designers and policymakers with options to support desired EBED outcomes, as discussed in Chap. 3 on the EBED process. Table 4.1 at the end of this chapter summarizes the information covered in this chapter and provides a quick reference of policy options for EBED practitioners.

Before proceeding with a more detailed description of supportive public policies, it is important to note several things about policy as a platform to strengthen the potential for EBED. First, the role of policies varies by EBED activity. Not all EBED activities require public policy intervention, although, presently, the majority of them do.[2] For those efforts that do require EBED policy intervention, the policy instruments can play either a supporting or a primary role. For example, a business incubator may train entrepreneurs to develop energy-based innovations and do so without the support of government policy. In contrast, a government may not have supportive policies for specific incubators but may adopt policies that are friendly for business start-ups, such as angel or investor tax credits, to improve access to capital; these government incentives thus play a secondary, supportive, but still critical role for EBED. As an example of a primary role, EBED programs that rely on Clean Development Mechanism funding for renewable energy projects may not be viable without this climate-based policy.

---

[1] We intentionally omit market reform from this list. Market reform refers to the restructuring of electricity markets in an effort to increase competition, market participants, and innovation. Energy market reform may entail the conversion of the power sector to a fully competitive market, in which only private utilities provide electricity. Market reform could alternatively entail simply involving private participants in the power sector or implementing policies that target management efficiency. Studies have found that some form of private participation in the electricity market has the potential to reduce hidden costs of electricity (Eberhard et al. 2010). Complete privatization, while once considered an important condition for fully functioning and efficient markets in both developed and developing countries, has been found to not always be the most effective or efficient solution for many countries (Blumsack et al. 2006; Wamukonya 2003; Williams and Ganadan 2006) and may in some cases negatively affect the quality and affordability of electricity service. In the interest of not delving too deeply in this ongoing debate, we do not include market reform in the list of EBED policy categories.

[2] All EBED initiatives, however, involve some degree of policy, if defined loosely, because they require a plan, statement of intent, procedure, or protocol. This chapter focuses specifically on public policies, that is, policies that involve public decision making.

4 Supportive Policies for Energy-Based Economic Development

Table 4.1 EBED policies

Policy category	Technology innovation	Technology adoption	Entrepreneurship	Industrial growth	Workforce training	Carbon and environment	Planning
Policy objective or focus	Encourage new and cutting-edge technology ideas	Assist in creating markets for technologies	Encourage the development and advancement of entrepreneurial activities	Support industry development and expansion	Increase opportunities in the labor force and connect people to jobs	Reduce environmental pollution and other environmental externalities	Prepare for future conditions and set goals
Types of policy instruments	-RD&D	-Feed-in tariffs	-Access to capital	-Business development	-Incentives	-Subsidies	-Integrated resource planning
	-Contracts	-Net metering and interconnection standards	-Infrastructure development	-Information and coordination	-Direct program provision	-Technology standards	-Comprehensive or strategic planning
	-Grants	-Loan guarantees	-Information and training	-Tax incentives	-Information provision	-Information	-Low emission planning
	-Tax credits	-Subsidies and other incentives		-Other incentives	-Partnership facilitation	-Education	-Sustainable cities planning
	-Inducement prizes	-Government procurement and demonstration		-Direct regulation		-Emission performance standards	
	-Public-sector research	-Information and education		-Import substitution		-Direct emissions regulation	
	Technical education and training	-Regulatory standards		-Export promotion		-Cap and trade	
	Intellectual property protection			-Foreign direct investment		-Taxes	

Second, the discussion in this chapter is not intended to identify policy winners nor sets of policies that are universally most appropriate. Rather, we note that a supportive policy environment can make a difference in EBED effectiveness. The projects and programs discussed in Chaps. 6 through 9 generally involve an active policy environment. Therefore, we believe that EBED practitioners and researchers should be aware of a range of EBED policies that can support their initiatives. It is up to the representative stakeholders to choose the policies most appropriate for their goals, objectives, assets, and chosen EBED approach.

Moreover, there is no one policy agenda for EBED activities, not only because EBED represents a range of different activities, but also because different places face a range of conditions, institutions, actors, and EBED pathways. There is also no set prescription for the number of policies that an initiative, or a country or locality, can or should adopt. Generally, it is the case that the more extensive the list of goals and objectives, the more likely it is that more than one policy is involved in supporting the initiative, each addressing distinct objectives. Top-down EBED approaches in particular tend to have multiple layers of EBED goals and objectives and, therefore, often have more than one policy. If one policy, however, is effectively able to target multiple objectives, then it is not necessary to adopt more than one policy to satisfy those multiple objectives.

Third, we do not analyze the effectiveness of these policies nor suggest ways that these policies can be made more effective. The policy literature highlights the need to design predictable and stable policies; align policies deliberately within broader policy portfolios; and carefully craft policy design features to maximize the effectiveness of the policy instrument. For additional insights on factors that improve policy effectiveness in the EBED realm, refer to Martinot et al. (2002), Geller (2003), Brass et al. (2012), Newell (2007), and Carley (2011).

Finally, similar to our discussion of EBED goals and approaches, the policy categories discussed below are not mutually exclusive. A government may develop an industry-specific policy at the same time that it shapes policy tools to facilitate technology innovation investments, for instance. Similar policies can fit in multiple categories, such as incentives. Technology standards, for example, are included below as both a technology adoption policy and an environmental policy. If the objective is to help a specific technology reach market maturity, then the policy should be classified as an adoption policy. Alternatively, if the objective is to help mitigate environmental pollution through the use of an advanced technology, then it can be classified as an environmental policy. But the policy itself is the same, just classified as fitting within two different subcategories.

With this groundwork established, we now discuss EBED-supportive policies.

4.1 Technology Innovation Policies

Technology innovation policies help bring energy innovations to market. These kinds of policies nurture the early stages of a "technology launch pathway" for innovators and entrepreneurs (Weiss and Bonvillian 2009). The standard S-shaped

4.1 Technology Innovation Policies

Fig. 4.1 Technology market transformation. *Source* Adapted and modified from Nadel and Latham (1998)

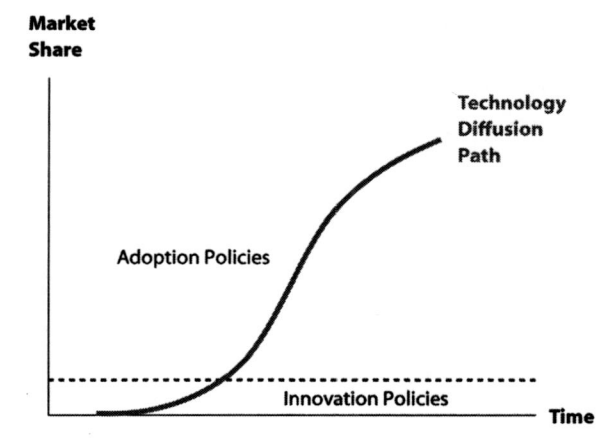

technology diffusion curve in Fig. 4.1 illustrates this pathway. This curve represents a general market transformation process for a typical technology, where the technology starts as a concept, becomes a proven concept, is introduced into the market, and evolves into mass-market adoption. Different policies play different roles through this diffusion process. Those policies that we classify as "innovation policies"[3] are most appropriate for technology development below the dashed line. We discuss in the next section "adoption policies," which generally occur above the dashed line.

Policies within the technology innovation category support the early stages of technology development, from idea creation through proof of concept. The primary policy under this category is research and development, or "R&D." Many also add a second "D" to this acronym to include deployment or demonstration of innovations.

R&D tends to include financial support for research contracts and grants, research tax credits for the financial sector, and innovation inducement prizes, such as those allocated by the Advanced Research Projects Agency-Energy (ARPA-E) in the United States. The government may also provide public-sector research through national research laboratories or pursue joint industry-government efforts. The Indian Council on Scientific and Industrial Research (CSIR) offers such an example. It was established in 1942 to promote scientific research, set up R&D institutes, and disseminate data on research and industry. Over the years, it has evolved to support RD&D for target growth and import-substitution industries. Today, CSIR has a staff of over 13,000 with a government-supported grant budget of $325 million (U.S. dollars). CSIR has increased income from outside sources

---

[3] As noted in Chap. 2, we recognize that innovation policies are not limited to technology development or technology adoption. Instead they emphasize the development or adoption of *something new*, including technologies but also business models and services, that can improve the welfare of the general public or spur economic opportunity.

from 1.8 billion rupees in 1995–1996 to 3.1 billion rupees in 2005–2006, approximately $65 million (World Bank 2010).

Another technology innovation policy is support for technical education and training, such as grants or other incentives offered to institutions that educate scientists within potentially promising innovative realms (see Romer 2001 for a more extensive discussion of subsidies for education and training). Finally, the government may provide intellectual property protection.

In many developing countries, technology innovation generally stems from technology transfer rather than R&D and "effective technology transfer is paramount for their innovation strategy" (World Bank 2010, p. 112). Policies supportive of a healthy business climate tend to simultaneously foster innovation. Policies supporting an attractive business climate encourage foreign direct investment and thus lure innovation in the form of foreign enterprise to locate in a developing country. Once these firms are in place, business climate policies help countries take advantage of the technology innovation spillovers by augmenting the country's ability to learn, adapt, and disseminate these innovations throughout their region (World Bank 2010, p. 116). This form of technology transfer makes it easier for local firms to adapt the technology for wider applications suitable for the local economy.

4.2 Technology Adoption and Commercialization Policies

Policies within the second category target technological adoption and thus aim to accelerate and enable EBED goals. These kinds of policies encompass a range of instruments that target power producers, energy system owners, and electricity and other energy users. Within the EBED framework, these policies are critical for encouraging consumers to use low-emissions, efficient, and advanced energy sources and techniques. Many countries, for example, are encouraging residents to adopt electric or hybrid vehicles to reduce the country's dependence on oil and lower carbon emissions.

We highlight the following policy instruments: feed-in tariffs, net metering, interconnection standards, framework laws, loan guarantees, incentives, government procurement and demonstration, information and education, and regulatory standards.

4.2.1 Feed-In Tariffs

Feed-in tariffs help reduce the cost and improve the investment recovery conditions of renewable energy. Many renewables have higher upfront costs than conventional sources, on a per-capacity basis, even though the costs of fuel are often much less, if not negligible (e.g., the wind does not cost anything). As a result, it is difficult for renewable energy producers to compete with conventional power producers, which introduces uncertainty about whether renewable energy investors will be able to secure an adequate return on their investments. These conditions make investment in renewable energy more risky.

Many countries have adopted feed-in tariffs to address this pricing barrier. Feed-in tariffs provide alternative power producers with a preferential price per unit of generation over a specified number of years. This price is often based on the average cost of generation for that resource, and rates often decrease over time. Feed-in tariffs also offer guaranteed, long-term contracts to renewable energy producers, which boosts certainty in the market and thus reduces risk for investors.

Over the past several decades, feed-in tariffs have become quite popular. As of 2012, feed-in tariffs were the most common renewable energy policy support instrument across the world. They are used in Germany, China, Israel, Thailand, Australia, India, Italy, and Uganda, among others.

4.2.2 Net Metering, Interconnection Standards, and Framework Laws

Net metering policies improve access to the electric grid for smaller, independent energy providers that own low-emissions energy systems, which is important for the scalability of EBED projects because it increases the number of potential energy providers in a place. Net metering requires that all small-scale energy system (referred to as "distributed generation") owners have access to the electric grid and that they can give or take electricity from the grid as needed. Net metering customers (i.e., those that own their own distributed generation units) can generally run their meter backward when they have excess electricity to put electricity back onto the grid. These system owners are then compensated for the electricity that they dispatched at the retail or wholesale price of electricity or some other rate depending on the structure of the net metering policy. Net metering policies are generally accompanied by interconnection standards, which specify the procedures that one must follow in order to hook their distributed generation unit up to the electric grid, including fees, legal requirements, paperwork, and contracts.

A net metering policy resembles a feed-in tariff in terms of the manner in which it provides financial compensation to an energy system owner, although the compensation is generally much higher for a feed-in tariff than it is for a net metering policy. At its core, however, a net metering policy ensures independent power producers access to the electric grid. Although many countries do not call their policies "net metering," a number of countries have regulatory frameworks for renewable energy, which allow independent power producers to connect their systems to the electric grid and generate and sell their electricity to utilities through power purchase agreements (Martinot et al. 2002). A growing number of developing countries in particular have adopted this type of regulatory framework over the past several decades.

4.2.3 Loan Guarantees

Loan guarantees are financing tools used to accelerate the pace at which that EBED-related inventions and businesses can reach target markets. Investors in emerging technologies and businesses that are close to commercial viability may not be able to secure loans at adequate rates in private credit markets because of a lack of market familiarity with the technology and the uncertainty of default. One mechanism for addressing this market barrier is for the government to provide a loan guarantee, which is a shared government-industry responsibility for a certain portion of a loan in the event that a debtor defaults (Newell 2007). The addition of government support and security gives developers access to lower-cost capital, because the perceived riskiness of the investment is reduced. Loan guarantees are particularly helpful to target independent power producers in electricity markets, as well as start-up companies that may not already have a proven investment track record and strong credit. Loan guarantees may also be used simply to support the growth of targeted industries, regardless of whether this industry has access to alternative capital. Loan guarantee programs are used across a variety of contexts. The United States, for example, provided approximately $3.9 million in funding for credit subsidies to back DOE loan guarantees for renewable energy, biofuels, and electric transmission upgrades through the ARRA.

4.2.4 Incentives

Governments may provide a variety of other incentives as well for energy or energy-related technology adoption, including tax incentives, grants, rebates, low-interest loans, and credit. Tax incentives provide a reduction in personal, corporate, sales, or property tax. An example of a tax credit in the U.S. context is the production tax credit, which was first passed in the Energy Policy Act of 1992 and subsequently reinstated on several occasions through other legislation. It provides 1.5 cents per kW h, adjusted for inflation and currently at 1.1–2.3 cents/kW h, for renewable energy generation companies for the first 10 years of system operation; eligible technologies include wind, closed- and open-loop biomass, geothermal, landfill gas, municipal solid waste, qualified hydroelectric, and marine and hydrokinetic energy of 150 kW or higher. Tax credits are less common in developing countries than they are in developed countries. Some developing countries, such as India, have experimented with investment tax credits for renewable energy with some positive returns (Martinot et al. 2002).

Low-interest loans are offered in many places around the world to target a variety of different market participants. The government may offer low-interest loans to entities such as public institutions or nongovernmental organizations for the purchase of renewable energy systems or energy efficient technologies. Many countries also offer low-interest loans to households that purchase their own energy applications or systems, such as solar lanterns, solar home systems, or efficient appliances. Governments may also provide loans to businesses and entrepreneurs that sell energy technologies to individuals or households, such as a solar photovoltaic rental company in the Dominican Republic that received a loan to purchase the solar panels and then charged homeowners approximately $20 per month to rent the panels (Martinot et al. 2002; Geller 2003). Some loans are used to support businesses or other entities that provide energy technologies and are able to offer credit to system owners. Grameen Shakti, a nonprofit vendor in Bangladesh, is a well-cited example of an organization that provides households with credit lines—for up to approximately 3 years with a down-payment of 15–20 % of the upfront cost—for solar home systems (Martinot et al. 2002). In other locations, national or local banks provide the line of credit for consumers who purchase their energy systems or technologies from specific, reputable companies.

Government incentives may also take the form of special "perks," or perhaps more accurately termed behavioral incentives. Examples include special access to high occupancy vehicle lanes for those drivers that have hybrid, electric, or another type of alternative vehicle; designated parking spots for alternative vehicles; free parking; or other preferential treatment for using a more efficient or low-carbon technology.

4.2.5 Government Procurement and Demonstration

In contrast to the government incentivizing other entities to adopt new technologies, it could instead set an example and purchase the technology itself. This type of policy is referred to as government procurement. The purchasing power of a government is relatively large, thus giving it power to scale adoption of EBED products in the market.

There are three types of innovation procurement policies: purchase of innovative goods and services, precommercial procurement, and catalytic procurement. Purchase policies are when government specifies that it is looking for a new or alternative solution to a public need. For example, a local government could state a need to convert their vehicle fleet to one that uses lower-carbon sources. Firms that produce electric, natural gas, biofuel, or hybrid vehicles may then submit their vehicles as an option for the government to purchase and use as its solution. Precommercial procurement is when R&D is still needed for certain technologies, and the scale of a government purchase is used to help advance the R&D to push the technology to market. In this instance, governments have to make their medium- to long-term needs known but have the patience for the product to develop to meet those specified needs. Japan and the United States, for example, have lowered the costs of fuel cell stations by supporting fuel cell-powered buses as a means of energy efficient public transportation (World Bank 2010, p. 127). Catalytic procurement is aimed at market transformation. A government could mandate that all new public buildings or upgrades must adhere to specific efficiency standards or use specific technologies, materials, or equipment. Or a government could purchase a bulk order of a specific technology to help establish the market for that technology.

Public procurement has the potential to advance significant technological development and adoption. As just one example, Edler and Georghiou (2007) found that between 1984 and 1998 48 % of Finland's successfully commercialized products resulted from government procurement policies.

4.2.6 Information and Education

Information and educational policies educate consumers through audits, labels, public information, advertisements, and professional and technical training about EBED options in the marketplace. As an example, in the United States several scholars have noted that consumers are less likely to purchase fuel-efficient vehicles because consumers do not fully account for the value of future fuel economy savings. In response to this problem, the U.S. Environmental Protection Agency and the National Highway Traffic Safety Administration joined forces to revise the labels that are placed on all new vehicles to more effectively relay this information to car buyers. All vehicles of model year 2013 and later will be required to display a new label that includes highway and city miles-per-gallon ratings, 5-year estimates of fuel savings compared with a conventional vehicle, estimates of how much fuel or electricity the vehicle requires to drive 100 miles, and smog and pollution emissions ratings. Driving range and charging times are also included for electric vehicles (U.S. EPA 2013).

On the training side, governments can provide technical maintenance and operations training to individuals within a community that installs distributed generation, such as solar photovoltaic panels, or other small-scale energy systems, such as cookstoves. In remote places where technical service providers are not readily available, training individuals within the community on maintaining their energy systems helps ensure the longer-term viability of these systems (Brass et al. 2012) and, in the context of this book, the EBED initiative.

4.2.7 Regulatory Standards

Regulatory standards, such as technology or portfolio standards, stipulate specific benchmarks that must be achieved in energy markets so that low-emissions technologies and energy sources can be more readily adopted in the marketplace. One of the most common standards used to date for EBED is the renewable portfolio standard, also referred to as a quota system, which is a mandate that a percentage of total electricity will come from renewable or alternative energy by a specific date (e.g., 20 % renewable electricity by 2025). These mandates are satisfied on an annual basis through the exchange of renewable energy credits or certificates, which represent a unit of generation, usually 1 MW h. This type of policy has been adopted by a number of countries and U.S. states in an effort to diversify these countries' electricity portfolios. Fuel standards represent a similar concept by which a locale must increase the amount of alternative fuel it uses by a certain time. Along the same lines, technology standards require the use of specific technologies to reduce carbon or other emissions at power plants, for example.

Efficiency standards are another type of regulatory standard; it requires a minimum efficiency level for all eligible technologies. Fuel economy standards are an example of an efficiency standard; a government sets a specified miles-per-gallon level that must at least be met, on average, across a manufacturer's vehicle fleet. The government could also set efficiency standards for appliances, equipment, or power plants. In the 1990s, South Korea, for example, set efficiency standards for appliances and lamps; the Philippines for air conditioners and refrigerators; and Mexico on automobiles, refrigerators, and air conditioners (Geller 2003). A vast number of countries have followed suit in more recent years and adopted similar, if not more advanced, standards. Some countries also set efficiency standards for the electricity sector; all utilities that operate within the sector must achieve certain levels of energy savings each year.

4.3 Entrepreneurship Policies

Entrepreneurship has increasingly become recognized as an important driver of growth in developed and developing economies. For EBED, entrepreneurship policies help disseminate low-emissions, efficient, and advanced products or methods into new markets while simultaneously creating new jobs and business opportunities. Policies to support entrepreneurship are implemented at multiple levels of government, from local to national levels, and aim to create an enabling environment that reduces costs and other barriers so that small to medium enterprises (SMEs) and individuals with ideas can succeed in the marketplace (Gilbert et al. 2004; Audretsch 2004; Lichtenstein and Lyons 2010). Entrepreneurship policies facilitate new firm creation, often in new and competitive sectors, and thus help seed a new set of competitive industries (Vivarelli 2012).

Entrepreneurs are defined by Lichtenstein and Lyons (2010) as "individuals who create value or wealth (i.e., business assets) by identifying and capturing market opportunities" (p. 33). Some entrepreneurs are driven by opportunity and innovation and seek to exploit new knowledge for commercial benefit in the marketplace; and some entrepreneurs are driven by necessity and become entrepreneurs because they need to survive. This latter type of entrepreneurship emerges in places with deep poverty and lack of formal employment opportunities. The informal sector, therefore, is where entrepreneurship of this kind emerges, typically with street vending and traditional and personal services (Vivarelli 2012,

p. 1). It is important to point out these distinctions because policies that support the former type of entrepreneur can often be unsupportive or even harmful to the latter type of entrepreneur, who may benefit more from antipoverty and skills-training policies (Karnani 2009).

Audretsch (2004) identifies another important distinction between entrepreneurship policies targeted toward SMEs and those aimed at entrepreneurs. SME policies are typically implemented by a government entity charged with supporting the growth of SMEs with 50–250 or fewer employees. Policies targeted to entrepreneurs are more wide ranging and seek to nurture a culture that brings ideas to market in a holistic and supportive manner. In this case, the government targets not only SMEs and entrepreneurs, but also individuals with potential to be entrepreneurs. In other words, entrepreneurship policy focuses on "the process of change," and policies that target SMEs focus on enterprise development (Audretsch 2004, p. 182). EBED programs can emphasize all types of entrepreneurship policies.

We summarize entrepreneurship policies into three categories: start-up and expansion capital, access to infrastructure and services, and awareness and training.

4.3.1 Start-Up and Expansion Capital

A common barrier to entrepreneurial growth is a lack of capital to finance the development and growth of an enterprise. Entrepreneurship is inherently risky because it involves the development of a new business model or alternative enterprise that is not necessarily proven in the marketplace. Thus, it is difficult for innovators to attract start-up capital for innovations, such as a new solar lantern technology or cocoa peat water filters. Many entrepreneurship policies, therefore, focus on addressing gaps in start-up and expansion capital.

Policies range from access to loan financing to support of risk capital, such as angel investors and venture capital in the early stages of business growth and equity capital in later stages of growth. Many countries and states offer tax breaks for individuals who invest in start-up companies. Since the 2007 to 2009 recession and collapse of financial institutions in the United States, for example, the emphasis on angels as investors in entrepreneurship has increased. Today, nearly half of the U.S. states have angel tax credit policies. An impact study conducted in the state of Wisconsin found that 930 full-time jobs were created within 125 companies that benefited from angel investments through the state's tax credit, with an average annual salary across these jobs of $83,000 (*Wall Street Journal* 2012).

A suite of microenterprise policies also supports smaller-scale enterprises in both developing and developed economies. As the term indicates, these policies support micro-loans and micro-credit schemes tailored to support financing needs that tend to be too small, too risky, and sometimes out of the service area for the traditional banking sector to service.

4.3.2 Access to Infrastructure and Services

Other policies that support entrepreneurs target ways to reduce the costs of infrastructure and support services associated with starting or growing a business. Managed or shared workspaces, for example, provided by business incubators and accelerators are supported through policy incentives. Legal, financial, and industry-specific professional advice are often coupled with the shared physical space to provide additional support to entrepreneurs to help them navigate the legal and financial complexities they will likely encounter as they grow.

Singapore offers an example of a country that uses policies supportive of business incubation to make it a hub of national and international entrepreneurship. Singapore supports entrepreneurial business incubation by providing shared space and professional support services. This service is not just provided for Singaporean entrepreneurs, but the economic development board there also uses this incubator-based support infrastructure to attract other incubators and entrepreneurs from around the world and thus create a global hub for entrepreneurship. From 2001 to 2005, business incubators and accelerators increased from 37 to 101, including 35 foreign incubators from the Asia Pacific region, Europe, United States, Dubai, and Saudi Arabia. These trends are making Singapore a hotbed for idea sharing, deal making, and entrepreneurship (World Bank 2010, pp. 88–89). A hub like this with a low-emissions energy focus would facilitate not only the creation of low-emission innovations, but also increase the number of entrepreneurs entering the market with products offering low-emissions alternatives.

Governments also support entrepreneurs with information on emerging technologies, competitive products, and markets. In-depth knowledge on these topics requires a nuanced and complex understanding of emerging markets, which makes it difficult for a small firm with limited resources to obtain the knowledge in absence of government assistance. To help these entrepreneurs overcome these information barriers, governments either have industry specialists serve as liaisons for entrepreneurs or governments facilitate connections between universities and relevant knowledge-based institutions and small businesses. The Netherlands, for example, offers SMEs knowledge vouchers. These vouchers give entrepreneurs free access to knowledge-intensive organizations such as public or private R&D facilities or industry specific experts (OECD 2007, in World Bank 2010).

4.3.3 Entrepreneurship Awareness and Training

A third category of entrepreneurship policies includes activities that raise awareness about entrepreneurship in general, provides skills training for idea generation and execution, and provides education and training on owning and managing a small business. Japan recognized the need for this kind of training for SME owners as early as 1963, when government officials began offering training for owners and managers of small firms through the Japan Small Business Corporation (Audretsch 2004). Policies supported through these kinds of efforts are also implemented to help reduce the administrative and bureaucratic burdens for small businesses or help them navigate government-related regulations more easily.

Policies that support entrepreneurship education aim to seed an entrepreneurial culture in communities, schools, and universities. Entrepreneurship training classes are being offered around the world in middle and high schools. Some colleges offer certificates and minors specifically in entrepreneurship. These types of programs tend to teach not only the technical aspects of operating a business, but also problem solving, creativity, social and interpersonal skills, continuous and independent learning, and innovation management (World Bank 2010).

4.4 Industrial Growth Policies

Industrial growth policies aim to complement or distort markets to strengthen an industrial sector or industry at large (Rodrik 2007, p. 100) or to "alter the sectoral structure of production toward sectors that are expected to offer better prospects for economic growth than would occur in the absence of such intervention" (Pack and Saggi 2006, p. 2).[4] Industry growth efforts typically aim to reduce costs, improve information exchange, and facilitate and enhance coordination between businesses within an industry and between business and government. The surge of economic growth experienced by Southeast Asian countries is well documented and often cited as examples of countries that effectively pursued industry growth policies, whereas countries in Latin America and Africa have not experienced similar results from such strategies (Wade 1990; Amsden 2001, 2007; Reinert 2008; Pack and Saggi 2006).

Industrial policies for EBED are used to support advanced, efficient, and low-emissions energy sectors for development and growth. Jacobs and De Man categorize industry policy in three ways: backing losers, picking winners, and building on existing strengths. Backing losers refers to government efforts to develop and strengthen declining industries. These industries tend to be important to the existing employment base but are often under threat from other regions that have become more competitive in the sector.

Picking winners refers to government efforts to target new and emerging industry sectors that policymakers anticipate will grow competitively over time and, therefore, offer a new source of jobs and revenue. The investments of countries and regions in biotechnology in the 1980s are an example of industry growth strategies that "picked winners." Many of the stimulus efforts in the recent

---

[4] There are differences of opinion regarding the definition for an industry growth strategy or industrial policy. The concept is, in fact, controversial in some settings. Pack and Saggi (2006) note that "[f]ew phrases elicit such strong reactions from economists and policymakers as *industrial policy*" (Italics in original, p. 2).

2007 recession included efforts to support targeted industries. Australia, the United States, South Africa, and China, in particular, targeted low-emissions industries to invest in for growth.

The third approach, building on existing strengths, is a blend of the first two. These policies target ways to accelerate growth within business activities that have already proved to be competitive in the marketplace. Brazil's economic pursuits in the agriculture and ethanol industry sectors are an example of an industry strategy that builds off a country's assets in sectors that are expanding global competitiveness (Rother 2010).

Related to industry growth policies are industry cluster policies. Industry cluster strategies were brought to the forefront of economic development practice with Michael Porter's *Competitive Advantage of Nations* (2000). These policies tend to focus on one or several industry sectors or a set of related competitive firms, rather than a "big push" for many sectors (Rodrik 2007). They also tend to be more subnational or localized in scope, whereas industrial policies are more national in scale (Porter 2000). Given that EBED is both national and subnational in reach, we highlight these distinctions at a high level to draw relevant points from the literature from both perspectives. Cluster policies can be either locally oriented, whereby policies aim to foster dynamic economies by removing market barriers and facilitating growth in a specific locale or "top-down," such as when national governments in conjunction with industry set national priorities and policies to support broader goals (Roelandt et al. 2000, p. 17).

As Spencer and his colleagues (2010) note, "Few constructs have enjoyed as much currency amongst both scholars and practitioners of regional economic development as the concept of the cluster" (p. 698). Cluster-specific approaches are distinct in their comprehensive nature to foster a specific value chain with targeted policy interventions on the supply and demand sides. As another scholar notes, "It is not the individual policy initiatives that are particularly unique, but rather the way in which they are formulated and targeted; that is in a manner that establishes—and mutually reinforces—the conditions for growth and development of key or promising end-market sectors and their supporting industries" (Feser 1998, p. 6).

Given the wide-ranging nature of industry growth policies, it is understandable that specific policies take many different forms, depending on the nature of the economy, scale, and scope of the intervention; the culture of institutions involved; and traditions of policymaking (Roelandt et al. 2000, p. 19). Further, there are many ways to characterize the menu of industry growth-related policies (Porter 2000; Feser 1998; Pack and Saggi 2006; Jacobs and De Man 1996; Roelandt et al. 2000; Rodrick 2007; Eisinger 1988).[5] For the purposes of this book, we categorize these policies as business climate, information and coordination, export promotion and foreign direct investment, and R&D policies.

---

[5] Industrial policy is a significant field of study. These sources offer just a few examples from which to glean a stronger understanding of industry-related policies in developed and developing economies today.

4.4.1 Business Climate Policies

Policies that use tax and finance tools to reduce production costs and create a stable and predictable environment in which businesses can operate are often supportive of EBED. For example, tax policies can create a stable business environment for companies and financial backers to invest in the related business. Policies can include tax exemptions or abatements on land and capital purchases or improvements, tax credits for the use or purchases of targeted goods and services, loans for working capital and fixed assets, and equity investments or tax incentives for job creation or industrial investment (Eisinger 1988; Rodrik 2007). For instance, a government may develop an eco-industrial park such as Burnside Park in Halifax, Nova Scotia, and partially build out its infrastructure at its own expense to have a "site-ready" facility in which businesses can locate. A government may elect to reduce the tax rate; defer taxes; or, in some instances, eliminate taxes for this industry with the agreement that this industry employ local workers and stay in the region for a certain amount of time. Similarly, a government may provide below-market loans to a specific company or sector to help support its growth.[6]

Tariffs are another policy tool to support business climates. Tariffs raise the prices of goods from outside markets within industries that countries have targeted for growth, thus giving domestic companies that produce or distribute these products a competitive edge over foreign companies.

Regulatory policies are also frequently used for industry development because they reduce uncertainties in the business climate and encourage the development and adoption of products and services in emerging industries (Roelandt et al. 2000; Porter 2000). These policies have played a prominent role in EBED recently to promote renewable energy industries, such as solar and wind, across the world. An example of a regulatory policy that has been used in the EBED realm is "buy local" regulations, where all goods that are used for specific initiatives must be manufactured within a country's border. For example, many of the energy programs under the ARRA had "buy American" requirements, where the goods used in weatherization of houses or other initiatives had to be sourced from U.S.-based companies.

4.4.2 Information and Coordination Policies

Information and coordination policies support EBED by reducing the costs associated with learning about an EBED-related industry sector, about regional assets, and about the effort to align this information toward market growth. These policies include governmental efforts that

---

[6] Rodrik (2007) provides an excellent summary table of illustrative industrial policies for over 30 countries for additional reference (pp. 123–130).

- facilitate knowledge exchange through formal and informal networks and events;
- provide strategic information about the industry in the region, its market trajectories, and related innovation assets that can be leveraged (e.g., university research, industry research, financing opportunities, or export promotion programs);
- organize internal government or related entities in public–private partnerships to better align with target industry needs;
- stimulate stronger relationships between suppliers, developers, contractors, and workers; and
- cultivate and promote international networks (Jacobs and De Man 1996; Porter 2000; Roelandt et al. 2000; Rodrik 2007).

Examples of efforts to enhance information exchange include Austria's innovation research program Technology Information Policy consulting (TIP). TIP, which is led by the Austrian government and conducted by relevant economic research institutes in Austria, conducts economic and industry studies that provide information and recommendations on policies to support innovation and growth. These reports also highlight specific sectors that are important to the country's innovation-led economic growth (Peneder 1997).

4.4.3 Import Substitution, Export Promotion, and Foreign Direct Investment Policies

Import substitution, export promotion, and foreign direct investment all aim to develop and expand a country's industries. These three categories of efforts employ taxes, grants, and regulations by governments to support and stimulate industrial growth (Amsden 2007).

Import substitution policies help countries expand their industrial base to replace imports with domestic products and, therefore, retain GDP that otherwise would leave the country. The process includes identifying imported products and services on which a country spends a significant amount of money. The government then helps seed and grow businesses to produce these same goods and services in country so that imports can be replaced with domestically produced products. Countries thus develop an industry base and retain the related economic benefits in country. Tariffs and entry restrictions for international firms are examples of policy tools used to protect domestic firms from outside competition. Low-interest credit, loans, and other financing tools, as well as performance standards for domestic firms, are all policy instruments that can be used to help grow domestic-based industries (Amsden 2007, pp. 92–95).

The electronics industry in Taiwan is an example of an industry supported by import substitution policies. Taiwan used high import tariffs in the electronics industry to help promote growth of this sector during the 1960s and 1970s. These import tariffs forced competitors from other countries such as Japan to establish joint ventures with the Taiwanese and engage in a technology transfer that helped Taiwan's electronics sector flourish (Amsden 2007).

Export promotion policies are often used in conjunction with import substitution policies to help businesses and industries in a region advance their knowledge about outside markets and facilitate networks and connections within and between markets. Policy efforts for export promotion involve information collection and exchange about potential markets for exporting, assistance in navigating through policies and procedures, and the facilitation of trade shows and trade missions with other countries to strengthen market relationships. For example, a country with a concentration of energy efficient building material producers may host trade shows in rapidly growing markets undergoing vast residential and commercial construction.

FDI policies integrate domestic industries with international markets. Government efforts to increase FDI involve linking businesses in a country with global networks and markets to provide these businesses technological knowledge and professional management techniques. Efforts under this umbrella include recruiting multinational companies to locate in county and facilitating mergers and acquisitions with a multinational company. For example, a government can create a special economic zone, such as a duty-free zone, to make it easier for a multinational company to locate within its borders. Jordan has several special economic zones to promote different regional industry strengths and to attract FDI in these sectors through the use of tax-free and limited regulations for exporting.

4.4.4 R&D for Industrial Growth Policies

Similar to the technology development and commercialization policies described earlier in this chapter, R&D policies support the creation, development, and dissemination of new products and services that help economies become or remain dynamic by replacing maturing industries with new emerging growth sectors. Such policies support the perpetual cycle of "creative destruction" described by Schumpeter (1950) as the process of "incessantly revolutionizing the economic structure from within, incessantly destroying the old one, incessantly creating a new one" (p. 84). As Malizia and Feser (1999) summarize, "This process takes the form of a cycle when entrepreneurs … introduce innovations. If these innovations are successful, imitators follow, any existing monopoly ends, investment for related innovation increases, and the economy begins an upswing" (p. 211). Policies include investments in public or public–private research facilities, tax credits for the support of small business research, or early-stage capital investments, such as angel investment tax credits. Federal research labs or federally sponsored research programs aligned with industry needs also qualify, such as ARPA-E in the United States and other examples already discussed above.

4.5 Workforce Development

Workforce development policies are important to EBED because they help individuals retain or get new jobs in the energy sector. These programs tend to focus on individuals and how to enhance their potential in the workforce.

Many define workforce development as including priorities such as creating opportunities for individuals within the labor market, connecting people to jobs, improving regional competitiveness, and facilitating employer–employee engagement (Jacobs and Hawley 2009; Harper-Anderson 2008; Golith 2000). Golith (2000), for example, describes workforce training as including "substantial employer engagement, deep community connections, career advancement, integrative human service supports, contextual and industry-driven education and training, and the connective tissue of networks" (p. 342). A 2008 OECD report notes the evolving roles of workforce development in a globalized economy; it explains that workforce development includes "attracting and retaining talent to solving skill deficiencies, integrating immigrants, incorporating the disadvantages not only into jobs but also into the education and training system, improving the quality of the workplace, and enhancing the competitiveness of local firms" (Giguére 2008, p. 21). Regardless of its wide-ranging targets and objectives, workforce development policies are often "go-to" EBED policies in times of economic distress because they increase the chances that someone can find a job or increase their salary, so these policies have the potential to put money directly in a person's pocket.

Since approximately the 1990s, "customized local strategies" for economic development have increased with an expanded integration of workforce development into economic development. These trends have led to an increase in sector-specific workforce training programs and the formation of partnerships between workforce agencies, economic development stakeholders, and private industry (Harper-Anderson 2008).

Workforce development policies include three basic activities: provision of incentives, direct provision of programs, and partnership facilitation. Governments can provide grants and other incentives to educators and trainers, employers, or employees. For instance, the government could subsidize a specific training program on smart grid professions or could offer companies within a specific industry grants to send their employees to new energy-based economic development training programs. Alternatively, the government could provide employee training programs on EBED professions directly or provide other information on professional opportunities in the EBED field. Finally, the government may facilitate working partnerships among businesses, academia, training programs, energy planners, and economic development planners that coordinate workforce training programs or curriculum.

4.6 Climate and Environmental Policies

Climate and environmental policies target emissions and other pollutants. Many of the policy instruments used for this purpose resemble others that have been identified above: subsidies for low- or no-emissions technologies that directly replace more emissions-intensive technologies, technology standards, information, and education. A number of additional policy instruments help support the low-emissions aspect of EBED as well, including emissions performance standards, direct emissions regulation, and taxes or cap-and-trade programs. We focus our discussion on these three policy instruments because the others are reviewed in earlier sections.

4.6.1 Emission Performance Standards

Emission performance standards set a limit on emissions per unit of output, such as a unit of GHG emissions per kW h of electricity or per gallon of gasoline. These standards can be applied on a specific entity, such as a power plant, or on an industry at large. Standards such as these encourage innovation and adoption of low-emissions technologies and products. Emission standards are more commonly used for transportation-based emissions than for electricity-based emissions. The United States, for example, has an emissions performance standard as part of its fuel economy standard for the transport sector. Although the United States had a fuel economy limit, referred to as the corporate average fuel economy, since 1975, as adopted in the Energy Policy Conservation Act, it revised these standards in 2010 and added a GHG emission limit per vehicle, as monitored by U.S. EPA.

4.6.2 Direct Emissions Regulation

Instead of setting an emissions level per unit of output, a government can set direct emission levels. Many countries, for example, have ratified the Kyoto Protocol and committed to legally binding GHG emissions reduction limits. Most countries participating in Kyoto targets have agreed to reduce GHG emissions to 1990 levels, although a handful of countries have identified different years as their base year to which they will peg their GHG target.

GHG limits that are set through the Kyoto Protocol are an example of governments setting economy-wide targets. It is also common for countries or subnational governments to set emission limits on specific industries or types of businesses. For example, the government could set an upper threshold of allowable emissions for all participants in the electricity sector or for specific power plants. Similar limits can be set on other emissions and pollutants as well, such as nitrogen oxides (NO_x), sulfur dioxide (SO_2), mercury, and lead, among others. These policies are directly relevant to EBED because they encourage the use of low-emissions technologies and energy efficiency.

4.6.3 Taxes and Cap-and-Trade Programs

A more flexible, or often referred to as "market-based" approach, to regulating GHG emissions and other pollutants is through a cap-and-trade policy or tax. The economics literature often upholds market-based approaches to environmental externalities as the most efficient of policy instruments when the objective is to reduce GHG or other emissions, because this approach directly internalizes the cost of the social problem (e.g., the social cost of carbon emissions that are not accurately reflected in the marketplace). These two approaches are differentiated according to what is set:

- A tax sets the socially optimal *price of a good*, such as a $/ton of GHG emissions.
- A cap-and-trade policy sets the socially optimal *quantity*, such as the total tons of GHG emissions for a given locale.

Many regions and countries have adopted one of these two policy instruments. For example, the European Union has adopted a cap-and-trade program. The EU Emissions Trading Scheme, launched in 2005, sets a carbon emissions cap for all participating countries, which applies to several but not all European industries. Those companies that take part in the cap-and-trade program are allocated pollution permits, which they trade among each other, or bank and borrow over time, in order to reach European annual emissions standards. British Columbia adopted a carbon tax in 2007. The tax started out at a fairly low rate of $10 per ton of carbon and then climbed to $30 per ton by 2012.

4.7 Planning

Even though some would argue that planning is not a public policy instrument, we think planning is essential to consider from an EBED framework perspective. Regional and local planning is often a necessary step to achieve an EBED goal. When a government sets a goal, such as a reduction in GHG emissions, or plans to create green jobs, it sends a message about priorities to its citizens and businesses. Plans also often lead to future public policies; the plan helps establish a policy idea and spark subsequent discussions about policy options. A plan may also be an explicit target that a government aims to achieve, which the government can support through incentives and other policies described in this chapter.

We highlight four types of EBED planning: integrated resource planning, comprehensive plans, low emission development planning, and sustainable cities planning.

4.7.1 Integrated Resource Planning

Integrated resource planning is the process by which a planning authority identifies the most cost-effective mix of supply- and demand-side energy resources that can be used in a locale to satisfy its energy demand. This kind of planning is helpful in EBED because it includes demand-side measures as a resource, so it helps utility planners incorporate opportunities for energy efficiency in the same manner that they plan for supply-side developments. Several countries, including Brazil, India, and Sri Lanka, have undergone integrated resource planning processes and increased support for energy efficiency as a result (Geller 2003). The use of integrated resource planning is also increasing across the world, with African countries showing particular interest.

4.7.2 Comprehensive and Strategic Planning

A comprehensive or strategic plan is a plan that documents all goals, objectives, areas of priority, and specific targets that a locale aims to achieve within a set time period. These types of plans also often include a statement of issues and goals and a plan for adoption, implementation, and monitoring. Comprehensive plans tend to focus on the physical, social, and economic dimensions of a place. Somewhat akin to the EBED process in Chap. 3, comprehensive plans help stakeholders in a location prioritize and organize to meet agreed-upon common goals. In the United States, comprehensive plans are often the product of local governments or communities and include elements such as increased walkability of neighborhoods and communities; more inclusive zoning for new, local businesses; and increased energy efficiency in public spaces. Other countries, such as China and Indian frequently produce strategic national plans that outline national objectives over a certain time frame.

An example of a national strategic plan is China's 5-year plans. China regularly sets 5-year plans, which outline future directions and goals. For example, in its 12th 5-year plan, China mapped its priorities between 2011 and 2015 and identified several national goals such as improving the environment, enhancing energy efficiency, and increasing domestic production over exports. The plan set emissions and development goals and identified seven priority industries for growth. These industries included new energy, energy conservation and efficiency, clean energy vehicles, biotechnology, new materials, new information technology, and high-end equipment manufacturing. Related to the discussion on industry growth strategies China is currently devoting significant resources toward these seven industries (KPMG China 2011).

4.7.3 Low Emission Development Planning

Low emission development strategies (LEDS) first surfaced in 2008 at the United Nations Framework Convention on Climate Change as national efforts to help guide countries toward low-carbon economic development. LEDS take on many different forms depending on the country context and its priorities, but these plans ultimately seek to integrate economic development and climate change planning. LEDS often complement or build on existing sustainability strategies, climate change strategies, and technology needs assessments as a means to help determine policy, capacity building, and funding priorities (Clapp et al. 2010).

LEDS strategies are varied, but they are all marked by a similar process. They are similar to EBED because they rely on stakeholder engagement and the exploration of policies to support low-emission development. They are distinct from EBED, however, in that they incorporate economic and energy modeling to develop development scenarios for a country. This modeling helps inform decision-making, where decision-makers have a better sense of potential environmental and economic effects of their actions.

The LEDS process begins with stakeholder engagement to help structure the framework for the entire planning process, followed by an assessment of existing economic and environmental data. Next, economic and environmental modeling helps develop an understanding about future trends for development. These models first establish a "business as usual" scenario, which is then augmented with alternative scenarios that help decision-makers evaluate options for alternative growth paths. With this data-driven analysis, stakeholders reconvene to consider and prioritize their choices. Finally, the process concludes with policy implementation and monitoring and evaluation (Open Energy Info n.d.).

4.7.4 Sustainable Cities Planning

Planning for sustainable cities is a recent offshoot of comprehensive urban planning that aims to reconcile the rapid rates of urbanization with out-of-date or inadequate urban infrastructure and the increasing vulnerability of cities from the impacts of climate change, food insecurity, and resource depletion. Those who advocate for sustainable cities planning call for a modernization of urban planning to more proactively address contemporary urban challenges and the reconfiguration of planning systems and government offices to work toward a cohesive national urban agenda that crosses policy sectors. Sustainable cities planning involves comprehensive strategies that incorporate environmental protection with infrastructure and real estate development by using more sustainable approaches to public transportation, water use and recycling, and solid waste management. Sustainable cities planning also calls for urban planners to recognize the positive role of the informal social and economic sectors of society, rather than excluding them from urban strategies. For instance, targeted neighborhoods or slums that were not commissioned from the planning department, yet in reality exist, should be incorporated into plans rather than ignored or demolished (United Nations Human Settlements Programme 2009).

North Port Quay, Australia, offers an example of a sustainable port city. A densely planned development within the city expects to construct 10,000 homes, public facilities, and a large marina, 100 % of which will be powered by renewable energy sources. Through solar photovoltaic wind turbines and an all-electric transport system, the city expects to be the world's first carbon-free development (Went et al. 2008; Newman et al. 2009, in UN-Habitat 2009).

4.8 Conclusion

To help digest and appreciate the portfolio of policies that can support an EBED program we list them in Table 4.1. The first row of this table lists the seven categories, the second row identifies their basic objectives or targets of these policies, and the third row displays the policy instruments that are used within the respective policy category.

References

Amsden A (2001) The rise of "the rest". Oxford University Press, New York
Amsden A (2007) Escape from an empire. The MIT Press, Cambridge
Audretsch DB (2004) Sustaining innovation and growth: public support for entrepreneurship. Ind Innov 11:167–191
Blumsack SA, Lave LB, Apt J (2006) Lessons from the failure of U.S. electricity structuring. Electr J 19:15–32
Brass J, Carley S, MacLean L, Baldwin E (2012) Power for development: an analysis of on-the-ground experiences of distributed generation in the developing world. Annu Rev Environ Resour 37:107–136
Carley S (2011) The era of state energy policy innovation: a review of policy instruments. Rev Policy Res 28:265–294
Clapp C, Briner G, Karousakis K (2010) Low-emission development strategies (LEDS): technical, institutional and policy lessons. Organisation for Economic Co-Operation and Development and the International Energy Agency, Paris
Eberhard A, Foster V, Briceno-Garmendia C, Shkaratan M (2010) Power: catching up. In: Foster V, Briceno-Garmendia C (eds) Africa's infrastructure: a time for transformation. World Bank, Washington, DC, pp 181–202

References

Edler J, Georghiou L (2007) Public procurement and innovation: resurrecting the demand side. Res Policy 36:949–963

Eisinger PK (1988) The rise of the entrepreneurial state: state and local economic development policy in the United States. University of Wisconsin Press, Madison

Feser EJ (1998) Old and new theories of industry clusters. Selected works. works.bepress.com/edwardfeser/3/. Accessed 5 June 2013

Gallagher KS, Grubler A, Kuhl L, Nemet G, Wilson C (2012) The energy technology innovation system. Annu Rev Environ Resour 36:137–162

Geller H (2003) Energy revolution: policies for a sustainable future. Island Press, Washington, DC

Giguére S (2008) A broader agenda for workforce development. In: Giguére S (ed) More than just jobs: workforce development in a skills-based economy. OECD Publishing, Paris

Gilbert BA, Audretsch DB, McDougall PP (2004) The emergence of entrepreneurship policy. Small Bus Econ 22:313–323

Golith RP (2000) Learning from the field: economic growth and workforce development in the 1990s. Econ Dev Q 14:340–359

Harper-Anderson E (2008) Measuring the connection between workforce development and economic development: examining the role of sectors for local outcomes. Econ Dev Q 22:119–135

Jacobs D, De Man AP (1996) Clusters, industrial policy and firm strategy. Technol Anal Strateg 8:425–438

Jacobs RL, Hawley JD (2009) The emergence of "workforce development": definition, conceptual boundaries and implications. In: Maclean R, Wilson D (eds) International handbook of education for the changing world of work. Springer, Netherlands, pp 2537–2552

Karnani A (2009) Reducing poverty through employment. Working Papers (Faculty)—University of Michigan Business School

KPMG China (2011) China's 12th five-year plan: overview. http://www.kpmg.com/CN/en/IssuesAndInsights/ArticlesPublications/Publicationseries/5-years-plan/Documents/China-12th-Five-Year-Plan-Overview-201104.pdf. Accessed 23 May 2013

Lichtenstein G, Lyons TS (2010) Investing in entrepreneurs: a strategic approach for strengthening your regional and community economy. Praeger, Santa Barbara

Malizia EE, Feser E (1999) Understanding local economic development. Center for Urban Policy Research, Rutgers, The State University of New Jersey, New Brunswick

Martinot E, Chaurey A, Lew D, Moriera J, Wamukonya N (2002) Renewable energy markets in developing countries. Annu Rev Energy Environ 27:309–348

Nadel S, Latham L (1998) The role of market transformation strategies in achieving a more sustainable energy future. American Council for an Energy-Efficient Economy, Washington, DC

Newell RG (2007) Climate technology deployment policy. Resources for the Future, Washington, DC

Newman P, Beatley T, Boyer H (2009) Resilient cities: responding to peak oil and climate change. Island Press, Washington, DC

Open Energy Info (n.d.) Low emission development strategies (LEDS) gateway. http://en.openei.org/wiki/Low_Emission_Development_Strategies. Accessed 18 June 2013

Organisation for Economic Co-operation and Development (OECD) (2007) Higher education and regions: globally competitive and locally engaged. OECD, Paris

Pack H, Saggi K (2006) The case for industrial policy: a critical survey. World Bank Policy Research Working Paper 3839. World Bank, Development Research Group, Washington, DC

Peneder M (1997) Creating a coherent design for cluster analysis and related policies: the Austrian TIP experience. Paper presented at the OECD Workshop on Cluster Analysis and Cluster Based Policies, Amsterdam, 10–11 October 1997. http://www.oecd.org/austria/2098045.pdf Accessed 18 June 2013

Porter M (2000) Location, competition, and economic development: local clusters in a global economy. Econ Dev Q 14:15–34

Reinert E (2008) How rich countries got rich... and why poor countries stay poor. PublicAffairs, New York

Rodrik D (2007) One economics, many recipes. Princeton University Press, Princeton

Roelandt TJA, Gilsing VA, van Sinderen J (2000) New policies for the new economy: cluster-based innovation policy: international experiences. Paper presented at the 4th Annual EUNIP Conference, Tilburg, The Netherlands, 7–9 December 2000

Romer P (2001) Should the government subsidize supply or demand for scientists and engineers? In: Jaffe AB, Lerner J, Stern S (eds) Innovation policy and the economy, vol 1. MIT Press, Cambridge, MA, pp 221–252

Rother L (2010) Brazil on the rise. Palgrave MacMillan, New York

Schumpeter J (1950) Capitalism, socialism, and democracy. Harper, New York

Spencer GM, Vinodrai T, Gertler MS, Wolfe DA (2010) Do clusters make a difference? Defining and assessing their economic performance. Reg Stud 44:697–715

United Nations Human Settlements Programme (UN-Habitat) (2009) Planning sustainable cities: global report on human settlements 2009. http://www.unhabitat.org/pmss/listItemDetails.aspx?publicationID=2831. Accessed 18 June 2013

U.S. Environmental Protection Agency (EPA) (2013) A new generation of labels for a new generation of vehicles. http://www.epa.gov/carlabel/. Accessed 21 May 2013

Vivarelli M (2012) Drivers of entrepreneurship and post-entry performance: microeconomic evidence from advanced and developing countries. Policy Research Working Paper. World Bank

Wade R (1990) Governing the market: economic theory and the role of government in East Asian industrialization. Princeton University Press, Princeton

Wall Street Journal (2012) Should angel investors get tax credits to invest in small businesses? Wall Street Journal, March 19

Wamukonya N (2003) Power sector reform in developing countries: mismatched agendas. Energy Policy 31:1273–1289

Weiss C, Bonvillian WB (2009) Structuring an energy technology revolution. The MIT Press, Cambridge

Went A, Newman P, James W (2008) Renewable transport: how renewable energy and electric vehicles using vehicle to grid technology can make carbon free urban development. CUSP Discussion Paper 2008/1. Curtin University Sustainability Policy (CUSP) Institute, Perth

Williams J, Ganadan R (2006) Electricity reform in developing and transition countries: a reappraisal. Energy 31:815–844

World Bank (2010) Innovation policy: a guide for developing countries. World Bank, Washington, DC

Chapter 5
Evaluation and Metrics

Abstract With the increased prevalence of EBED-related initiatives, evaluators are increasingly attracted to this topic area, and the number of evaluations in this domain will likely increase. Evaluations offer information that generates a shared understanding of what works well and what does not in areas of practice. In particular, a collection of quality evaluations in this domain will facilitate a stronger community of research and practice in EBED. To help guide evaluations, this chapter provides a review of the state of EBED evaluations in the peer-reviewed literature and suggests how a stronger understanding of the holistic nature of EBED can strengthen evaluations in the future. The discussion in this chapter focuses on outcome measures, type of initiatives studied, methodological approaches, and timing and research design. This chapter also offers suggestions for evaluators to consider employing as the field advances.

Evaluation is a significant part of the EBED process because it allows stakeholders to assess how well their program or project conforms to their original goals, whether specific objectives were met, and how to improve or tailor their action plan accordingly. Evaluations are important to any program, but they are imperative for EBED because it is an evolving domain that does not have an extensive history of evaluation and, thus, lessons that can be translated from one project to another. Information distilled from evaluations offers critical insights to the broader EBED community on the relative effectiveness of different projects, processes, and techniques. This information, over time, generates a shared understanding of lessons learned for this practice area at large.

With the emergence of EBED-related initiatives, evaluators are increasingly attracted to this topic area, and we expect evaluation in this domain to become more prolific. To help guide these emerging evaluations, we provide a review of the state of EBED evaluations in the peer-reviewed literature and suggest how a stronger understanding of the holistic nature of EBED can strengthen evaluations in the future.

Inherent in EBED initiatives are a range of evaluation approaches. EBED can be targeted to places, markets, and households; and approaches occur at different geographic levels with different scales of transformation. Accordingly, EBED projects not only differ in their design, but they also range widely in their scale of investment. Although we cannot adequately cover this entire range or draw conclusions about the relative merits of one type of evaluation approach over another, it is possible to present an overview of EBED evaluation efforts, discuss themes within them, and offer suggestions for improving future evaluations.

This chapter builds off the discussion on metric identification, monitoring, and evaluation in Chap. 3. In Chap. 3, we highlight the need of program designers to incorporate appropriate evaluation metrics to track the initiative's operations and progress. In this chapter, we shift our focus to identify common practices as reflected in the peer-review literature so that we can expand on this foundation of understanding to offer ideas for enhancing evaluation efforts in the future. In this way, evaluations can more fully reflect the wider suite of EBED goals as this domain area continues to grow.

The discussion in this chapter rests on an extensive review of primarily peer-reviewed literature in the fields of energy, economics, development, and the environment.[1] The trends that we highlight below are based on this review and fall into four main areas: outcome measures, type of initiative, methodological approaches, and timing and research design. In each section below, we describe the findings from the literature, highlight how they are beneficial for the EBED field, and offer suggestions for evaluators to consider employing as the field advances.

5.1 Outcome Metrics

The first major trend in the literature involves the outcome measures most commonly studied. Metrics used in EBED evaluations generally include job estimates, but some studies also include income measures (e.g., GDP or household income), energy savings, or reductions in GHG emissions. On the whole, however, the focus of these analyses is on jobs created as a result of energy-related projects. For the analysts that rely on job-related metrics when assessing the outcomes of low-emissions or efficient energy projects, the measures often include "green jobs." We further describe "green jobs" here because it is a term that is often used without a clear definition.

The U.S. Bureau of Labor Statistics (n.d.) defines green jobs as involving one of the following conditions:

1. when a job is associated with a business that generates goods or services that have environmental or natural resource benefits,

---

[1] Five separate researchers in 2010 and 2012 conducted independent literature reviews and compared their lists of relevant literature. They read all articles identified to determine whether the effort under analysis was an EBED-related activity and whether the study presented quantitative outcome measures.

5.1 Outcome Metrics

2. when one's duties as part of their employment involves reducing the environmental footprint or resource extraction associated with production processes.

The definition of a green job, however, varies considerably across sources. Some include occupations that involve an environmental focus, and others include all occupations that produce environmental benefits (Bowen 2012). Some also narrow their definition to include only renewable energy occupations, while others include all environmental service and energy-related product occupations (Bowen 2012). The UNEP definition, for example, is very specific about the types of industries and occupations that are included; green jobs are defined as

> ... work in agricultural, manufacturing, research and development (R&D), administrative, and service activities that contribute [sic] substantially to preserving or restoring environmental quality. Specifically, but not exclusively, this includes jobs that help to protect ecosystems and biodiversity; reduce energy, materials, and water consumption through high-efficiency strategies; de-carbonize the economy; and minimize or altogether avoid generation of all forms of waste and pollution. (UNEP/International Labour Organization/International Organisation of Employers/International Trade Union Confederation 2008).

In the EBED realm, job estimates are generally reported as either total number of jobs per megawatt of energy production or as an aggregate number of jobs associated with energy projects. Studies that focus on EBED-related government spending, such as stimulus funding spent on energy development, tend to report total dollar or equivalent spent per job.

The methodology for measuring job changes, as well as the details as to what exactly is captured in the job estimate figure, varies significantly across studies. Although not necessarily the most common or the most widely accepted way to report job figures, the most precise way is to report the figure as a full-time equivalent job-year (Wei et al. 2010). This term and measurement allow for consistency in the estimates and do not mask distinctions between full-time and part-time jobs, for example, or between temporary and permanent jobs. Alternatives to the full-time job-year, including one or several of the following, are also present in the literature:

- total jobs overall
- full-time equivalent jobs
- part-time employment
- monthly or quarterly job equivalents
- jobs gained versus jobs lost

In addition, some analysts include only new jobs that arise from new activities, while others count previous employment that has a "green" component. Estimates also vary according to whether an analyst measures direct or indirect estimates of job growth or both. Direct estimates are jobs created specifically and directly by the initiative, policy, or intervention; indirect estimates account for jobs further down the supply chain or that spill over into other sectors as a result of an intervention.

Differences between these various metrics may seem minor, but it is important to note that they can yield substantially different results, making cross-project comparisons difficult at best and inaccurate at worst. Because analysts' interpretations of the green growth concept vary, studies offer substantial differences in green job estimates (Gülen 2011). The OECD (2010), for example, observes in its interim report on implementing a green growth strategy that the employment multipliers used in assessing job creation potential from green stimulus vary across studies and countries. One should, therefore, exercise caution in comparing potential economic development benefits across even seemingly identical projects, because the assumptions that are embedded in analyses may differ significantly.

Other outcome metrics analyzed in the EBED literature include gross domestic or state product, output multipliers, household income, saved GHG emissions, and saved expenditures. GDP is a macroeconomic indicator of the total effect on a state or national economy. Multipliers are estimates that measure the increase in economic output per dollar spent on an EBED project. Those that report estimates of saved GHG emissions are interested in the environmental effects of EBED projects as well.[2] Measures of expenditures generally include saved expenditures on imported energy or saved expenditures per megawatt of new renewable energy capacity. Table 5.1 summarizes EBED studies by the outcome metrics they evaluate. This table includes only studies that provide quantitative estimates of EBED-related projects or efforts. It does not include case studies of individual energy provision and efficiency projects in developing countries, such as those prepared and published by the World Bank or international donors. For more extensive details on each of these studies, refer to the literature table published in Carley et al. (2011).

In the developing country context, measures of jobs and, in particular, green jobs are less common. The EBED evaluation literature from this perspective tends to either focus on GDP when the initiative under evaluation is a large-scale, nationally driven effort or on energy-specific outcome measures when the initiative is a smaller-scale project. Brass et al. (2012) review the peer-reviewed case study literature on distributed generation[3] projects in developing countries, for which the goals of these EBED initiatives most often include a reduction in poverty and energy poverty but may also include energy diversification, innovation and entrepreneurship, or energy efficiency. They find that the majority of studies report one or several of the following project outcomes: number of energy units installed, percentage of households in a region that use the energy technology, cost comparisons across technological options, and household energy savings. Very few case studies evaluate longer-term outcomes, including both technical aspects of the

---

[2] It is important to note that studies commonly measure the GHG emissions from energy projects. We do not, however, include these studies in Table 5.1 if they do not have an explicit economic development component as well, as defined in either the project approach (refer to Chap. 3) or in the outcome metrics.

[3] Distributed generation refers to small-scale, localized energy systems, either at the household, business, or community level. Brass and her colleagues' study includes distributed generation projects at all three of these levels.

5.1 Outcome Metrics

Table 5.1 Measuring EBED outcomes

Assessment category	References
Jobs per energy output	Kammen et al. (2006), Moreno and Lopez (2008), Pedden (2006), Sastresa et al. (2010), Simons and Peterson (2001)
Jobs per energy output + other factors	Álvarez et al. (2009)[a], Blanco and Kjaer (2009), Singh and Fehrs (2001), Tourkolias and Mirasgedis (2011)
Total jobs	Algoso and Rusch (2004), Bezdek for the American Solar Energy Society (2007), Global Climate Network (2010), KEMA (2009), Lehr et al. (2008), Rutovitz (2010), Ulrich et al. (2012), Upadhyay and Pahuja (2010), Wei et al. (2010)
Total jobs + other factors	Allan et al. (2008), Caldés et al. (2009), del Rio and Mercedes (2009), Hillebrand et al. (2006), Laitner and McKinney (2008), Neuwahl et al. (2008), Reategui and Tegen (2008), Roland-Holst (2008), Simas and Pacca (2011), Slattery et al. (2011), Wicke et al. (2009), Williams et al. (2008)
GDP	Ragwitz et al. (2009)
Multipliers	MacGregor and Oppenheim (2008)

Source Table originally sourced from Carley et al. (2011) with several modifications and additions
[a] Lantz and Tegen (2009) review the methods and assumptions used in the Álvarez et al. (2009) paper and argue that the conclusions drawn are not supported by the analysis and that the authors do not rely on traditional and accepted empirical techniques to determine job estimates

energy systems and development outcomes such as health, education, or environmental impacts.

These reviews reveal helpful suggestions for examining EBED outcomes and also highlight gaps that future evaluations can address. There is opportunity for EBED evaluations to incorporate additional outcome dimensions beyond just economic and energy-related metrics. Social and environmental outcomes, for instance, can help capture and communicate the cross-disciplinary nature of EBED activities. As discussed above and as made evident in Table 5.1, social, environmental, and other such dimensions are far less common in the evaluation literature. In support of this point, Brass et al. (2012) find that measures of environmental impacts, educational effects, health outcomes, social and cultural changes, and political consequences are rarely measured in distributed generation case studies, despite the fact that these factors are often documented as guiding project objectives.

A study prepared by del Rio and Burguillo (2009) also identifies the lack of EBED analyses that estimate additional outcome figures beyond the standard set of economic development measures. In response to the expressed need for other outcome measures that can provide insights on the effects or effectiveness of EBED efforts, del Rio and Burguillo consider the following outcome criteria:

- measures of demographic, education, and other human capital
- tourism, social cohesion, and environmental impacts
- the degree to which local resources are used
- personal perceptions about benefits
- the degree to which local industries are created.

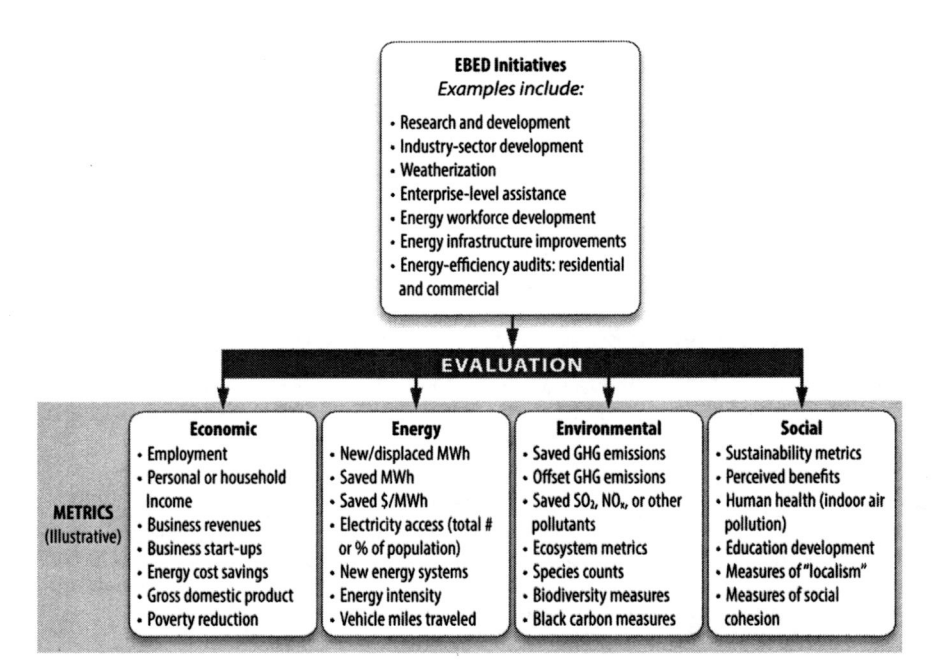

```
                    ┌─────────────────────────────────┐
                    │        EBED Initiatives          │
                    │        Examples include:         │
                    │                                   │
                    │ • Research and development       │
                    │ • Industry-sector development    │
                    │ • Weatherization                 │
                    │ • Enterprise-level assistance    │
                    │ • Energy workforce development   │
                    │ • Energy infrastructure improvements │
                    │ • Energy-efficiency audits: residential │
                    │   and commercial                 │
                    └─────────────────────────────────┘
                                    │
                    ▼▼▼ EVALUATION ▼▼▼
```

	Economic	Energy	Environmental	Social
METRICS (Illustrative)	• Employment • Personal or household Income • Business revenues • Business start-ups • Energy cost savings • Gross domestic product • Poverty reduction	• New/displaced MWh • Saved MWh • Saved $/MWh • Electricity access (total # or % of population) • New energy systems • Energy intensity • Vehicle miles traveled	• Saved GHG emissions • Offset GHG emissions • Saved SO_2, NO_x, or other pollutants • Ecosystem metrics • Species counts • Biodiversity measures • Black carbon measures	• Sustainability metrics • Perceived benefits • Human health (indoor air pollution) • Education development • Measures of "localism" • Measures of social cohesion

Fig. 5.1 EBED outcome measures. *Source* adapted from Carley et al. (2012)

Houser et al. (2009) also suggest a set of "green recovery metrics" for the evaluation of large-scale federal stimulus funding; metrics include speed of implementation, employment, energy savings, reduction of oil imports, and climate change measures.

Figure 5.1 illustrates the various outcomes that one could measure in an evaluation of EBED projects' effects or effectiveness. Of course, choices of outcome measures depend on the objective of the EBED effort, but there is room for EBED practitioners and researchers to embrace a wider set of metrics to account for the cross-disciplinary kinds of outcomes one can expect from EBED projects. The outcomes presented in this figure are not intended to serve as an exhaustive list but, rather, serve as a representative range of outcomes.

5.2 Type of Initiative Evaluated

The second main trend in the EBED evaluation literature is that analysts most commonly evaluate large-scale investments or policies related to energy supply and energy efficiency. It is less common for quantitative evaluations to focus on other EBED activities.

5.2 Type of Initiative Evaluated 87

Examples of job studies on large-scale energy-related investments are in a meta-analysis conducted by Wei et al. (2010). These authors found that renewable energy and other low-carbon energy-related sectors generate more jobs than fossil fuel-based sectors per unit of energy output. These analysts also ascertained that a U.S. energy blend with at least 50 % of its power sourced from renewable energy, energy efficiency, and other low-carbon sources such as nuclear and carbon capture and storage can generate over 4 million job-years between 2010 and 2030. Others have also studied job growth potential of new energy development and "green" growth and found similar benefits, although with significant variation in estimates and input modeling assumptions (GHK 2009; World Bank n.d.).

Studies that concentrate on energy use and demand-side energy efficiency tend to document the jobs associated with construction and installation and the estimates of cost savings from unspent energy budgets that can be redirected toward other sources of economic activity (Ronald-Holst 2008; MacGregor and Oppenheim 2008; Laitner and McKinney 2008). A common conclusion of many works in this field is that the creation and adoption of low-emissions energy technologies aimed at energy generation and efficiency tend to create more jobs per unit of installed capacity than conventional energy approaches or projects. In many of the studies that have been conducted in the United States, researchers note that a large percentage of employment from new, low-emissions energy and energy efficiency projects is jobs that are guaranteed to remain domestic (i.e., they are not at risk of being fulfilled by overseas labor) because the installation and construction of energy systems and technologies involves site-specific work (Center for American Progress 2009; Apollo Alliance 2009; Blue-Green Alliance 2009; Walsh and White 2008). Studies are less consistent in their conclusions about domestic versus international employment for energy systems' manufacturing components.

This trend shows that evaluators have been more prolific in publishing research that focuses on large-scale energy investments that have economic growth-related outcomes. We encourage analysts to also disseminate material on smaller projects at the city or state level, for example. Studies that focus on the specific unit of project intervention, such as the local community in which an EBED effort is applied and not more generally on the national scale, will contribute to additional insights for researchers and practitioners. This will open up the relevance of EBED studies to the broader evaluation community and shed light on the other kinds of approaches relevant for EBED.

5.3 Methodological Approach

The third trend within the literature involves the methods used to evaluate EBED initiatives. With the growing number of studies that measure the effects of EBED-related programs, most research employs economic modeling approaches, despite the fact that evaluation methods in this field vary. The modeling approaches in the literature tend to emphasize input–output techniques with a focus on jobs and other economic growth impacts.

Because these models are so predominant in the EBED evaluation literature, we describe them in greater detail. Input–output models capture the exchanges of goods and services between different sectors of the economy to determine overall employment changes and multipliers. These models often include a substantial amount of industry-related data and are thus quite large and time consuming to construct (Wei et al. 2010). Directly related to the discussions above, most EBED-related input–output studies in the literature seek to determine the economic impact from the growth in the energy supply and energy efficiency industry in the form of job creation (Wei et al. 2010; Lehr et al. 2008; Hillebrand et al. 2006; Ziegelmann et al. 2000). Scott et al. (2008) provide additional quantitative measures, such as energy saved, cost savings in investments in building stocks, and effect on national earnings. Within this economic impact literature, research typically focuses on the effects of specific policy interventions, such as those from solar thermal electricity deployment (Caldés et al. 2009) or wind energy projects (Williams et al. 2008).

Integrated approaches also surface in the literature. These are approaches with a combination of input–output and other economic models that determine effects on additional aspects of development, such as human capital and technological development (Sastresa et al. 2010) and local sustainability (del Rio and Burguillo 2009). Some supplement economic modeling with case studies to further describe the context of the economic impact (Bezdek 2007). Other methods to evaluate EBED include macroeconomic forecasting, general equilibrium models, regression analysis, analytical models using spreadsheets, and technical feasibility or other engineering analyses. The granularity of these approaches varies, as does the type of information that one can extract from the modeling outputs.

Aside from these quantitative modeling techniques, the evaluation literature focuses less extensively on other kinds of methods. A couple of noteworthy exceptions include logic models that help evaluators effectively track inputs, outputs, and outcomes of EBED efforts (Houser et al. 2009; Annecke 2008) and those who advocate for "development assistance criteria" to guide evaluations. The criteria for evaluation within this paradigm are relevance, efficiency, impact, and sustainability (Annecke 2008). All of these techniques introduce a qualitative component to the evaluation design. The logic model, for example, blends qualitative assessments with approaches that help evaluators calculate relevant outputs and outcomes consistently. Program analysts use the logic model to monitor the entire sequence of program elements, from goal formation through program outcomes. These alternative evaluation methods also allow for multiple dimensions of program success. They do not simply measure success as job growth, for instance. The methods can also be used ex post to monitor and evaluate actual program outcomes, rather than used to estimate hypothetical program outcomes, as is more common with a modeling tool such as an input–output model.

There are vast opportunities to expand and refine methodological approaches employed for EBED evaluations. Of course, it is important to tailor the methods to each project and, as discussed in Chap. 3 be deliberate in the choice of evaluation method so that measured outcomes match EBED goals and objectives. The choice of evaluation method will, and should, vary according to the project and the specific circumstances of the project location or community. Different types of EBED efforts will lend themselves better to different methodological approaches. A nationally driven or market approach, for example, may require the use of a sophisticated modeling effort, such as input–output, econometric, or generalized equilibrium modeling. Locally driven or smaller-scale projects may require an approach that is more context specific and has finer granularity, such as a logic model or basic analytical model.

For a cross-disciplinary initiative or policy, multiple evaluation approaches may be appropriate, depending on the scope of the project. The literature reveals that for EBED evaluations, multimodal approaches are less frequent. Because each method is limited in some way, however, employing more than one method may help EBED analysts and practitioners document a broader range of outcomes and impacts. We believe this is an important area for further investigation in the EBED realm.

5.4 Timing and Research Design

The final trend in the literature that we highlight is the timing of the EBED evaluation. The majority of the literature consists of modeling estimates performed ex ante or before the intervention occurs, rather than assessment of effects ex post or after implementation (see, for example, Algoso and Rusch 2004; Bezdek 2007; Hillebrand et al. 2006; KEMA 2009; Lehr et al. 2008; MacGregor and Oppenheim 2008; Moreno and Lopez 2008; Ragwitz et al. 2009; Singh and Fehrs 2001; Williams et al. 2008). Ex ante studies are important in that they help guide decisions about how to best allocate resources among different options. These studies also inform stakeholders and investors with information about expectations for project outcomes. On the other hand, ex post analyses provide quantitative and qualitative information about actual project effects. These studies are grounded in data collection about baseline conditions, the intervention, and outcomes from the activities that comprise the intervention. Evaluations conducted "after the fact" provide information that can be compared with preproject estimates to identify which factors contribute to project success and which do not.

The timing of the evaluation is also important to consider. Evaluation time horizons can range from short term to long term. Some authors distinguish between short, medium, and long; some also include the very short term. In general, short-term effects are those that occur directly after some form of intervention, such as a program, effort, or policy, when those involved respond directly to the change that is being evaluated. For example, direct employment effects from EBED intervention are a short-term effect. Medium-term effects are generally measured after the intervention has had enough time to diffuse through the target population, community, or economy. Long-term effects are after a period of time when the EBED intervention has affected technology decisions, industrial composition, and human behavior. Other analysts do not draw these distinctions and merely report an outcome measure that is not time specific.

Along these lines, a final proposed area for advancing the EBED evaluation framework is conducting, documenting, and publishing empirical research in a more comprehensive manner. This will help capture impacts along the complete project time horizon, control for intervening factors, and assess what about the process leads most significantly to successful outcomes. EBED program evaluations in particular, as distinguished from exploratory studies that take interest in the potential effects of an EBED effort, will be stronger if they employ a comparison calculation of pre- and postproject outcome measures. Additionally, empirical studies that measure which factors contribute most to the success of various EBED projects, and are able to control for other factors, will improve our collective understanding of how EBED projects work and whether they work as intended.

On the whole, moving forward in evaluative efforts, it will be helpful for the EBED community to build on the knowledge base of existing evaluation research and expand it to include a wider set of metrics and methods. We encourage additional research, discussion, and modification to the scope of EBED evaluations in the following areas:

- using additional metrics,
- expanding the types of initiatives that are evaluated and published,
- incorporating multiple methodological approaches that are suited for each type of evaluation,
- conducting evaluations during and after project implementation.

We encourage others to continue to build on this research to expand and refine evaluation to meet the growing number of EBED projects.

References

Algoso D, Rusch E (2004) Renewables work: job growth from renewable energy development in the Mid Atlantic. Public Interest Research Group, Trenton

Allan GJ, Bryden I, McGregor PG, Stallard T, Swales JK, Turner K, Wallace R (2008) Concurrent and legacy economic and environmental impacts from establishing a marine energy sector in Scotland. Energ Policy 36:2743–2753

Álvarez GC, Jara RM, Julián JRR, Bielsa JIG (2009) Study of the effects on employment of public aid to renewable energy sources. Universidad Rey Juan Carlos, Madrid

Annecke W (2008) Monitoring and evaluation of energy for development: the good, the bad and the questionable in M&E practice. Energ Policy 36:2839–2845

Apollo Alliance (2009) Make it in America: the apollo green manufacturing action plan. Apollo Alliance, San Francisco

Bezdek R (2007) Renewable energy and energy efficiency: economic drivers for the 21st century. American Solar Energy Society, Boulder

References

Blanco I, Kjaer C (2009) Wind at work: wind energy and job creation in the EU. European Wind Energy Association, Brussels

Blue-Green Alliance (2009) Building the clean energy assembly line: how renewable energy can revitalize U.S. manufacturing and the American middle class. Blue-Green Alliance, Minneapolis

Bowen A (2012) "Green" growth, "green" jobs and labor markets. World bank policy research working paper 5990. World bank, sustainable development network, Washington

Brass J, Carley S, MacLean L, Baldwin E (2012) Power for development: an analysis of on-the-ground experiences of distributed generation in the developing world. Annu Rev Env Resour 37:107–136

Bureau of Labor Statistics (n.d.) Green jobs. http://www.bls.gov/green/#overview. Accessed 23 April 2013

Caldés N, Varela M, Santamaria M, Sáez R (2009) Economic impact of solar thermal electricity deployment in Spain. Energ Policy 37:1628–1636

Carley S, Lawrence S, Brown A, Nourafshan A, Benami E (2011) Energy-based economic development. Renew Sust Energ Rev 15:282–295

Carley S, Brown A, Lawrence S (2012) Economic development and energy: from fad to a sustainable discipline? Econ Dev Q 26:111–123

Center for American Progress (2009) The clean energy investment agenda: a comprehensive approach to building the low-carbon economy. Center for American Progress, Washington

del Rio P, Burguillo M (2009) An empirical analysis of the impact of renewable energy deployment on local sustainability. Renew Sust Energ Rev 13:1314–1325

GHK (2009) The impacts of climate change on European employment and skills in the short to medium-term: a review of the literature. Final report to the European Commission Directorate for Employment, Social Affairs and Inclusion Restructuring Forum, vol 2. GHK, London

Global Climate Network (2010) Low-carbon jobs in an interconnected World. Global climate network discussion paper no. 3. Global Climate Network, London

Gülen G (2011) Defining, measuring and predicting green jobs. Copenhagen Consensus Center, Lowell

Hillebrand B, Buttermann HG, Behringer JM, Bleuel M (2006) The expansion of renewable energies and employment effects in Germany. Energ Policy 34:3484–3494

Houser T, Mohan S, Heilmayr R (2009) A green global recovery? Assessing U.S. economic stimulus and the prospects for international coordination. World Resources Institute Policy Brief, Washington

Kammen D, Kapadia K, Fripp M (2006) Putting renewables to work: how many jobs can the clean energy industry create?. University of California-Berkeley, Renewable and Appropriate Energy Laboratory (RAEL), Berkeley

KEMA (2009) The U.S. smart grid revolution: KEMA's perspectives for job creation. GridWise Alliance, Washington. www.gridwise.org/resources_gwaresources.asp. Accessed 11 Nov 2012

Laitner JA, McKinney V (2008) Positive returns: state energy efficiency analyses can inform U.S. energy policy assessments. American Council for an Energy Efficient Economy, Washington

Lantz E, Tegen S (2009) NREL Response to the report study of the effects on employment of public aid to renewable energy sources from King Juan Carlos University (Spain). National renewable energy laboratory. White paper NREL/TP-6A2-46261. Contract no. DE-AC36-08-GO28308. National Renewable Energy Laboratory, Golden

Lehr U, Nitsch J, Kratzat M, Lutz C, Edler D (2008) Renewable energy and employment in Germany. Energ Policy 36:108–117

MacGregor T, Oppenheim J (2008) Energy efficiency equals economic development: the economics of public utility system benefit funds. Entergy, New Orleans

Moreno B, Lopez AJ (2008) The effect of renewable energy on employment: the case of Asturias (Spain). Renew Sust Energ Rev 12:732–751

Neuwahl F, Löschel A, Monelli I, Delgado L (2008) Employment impacts of EU biofuels policy: combining bottom-up technology information and sectoral market simulations in an input–output framework. Ecol Econ 68:447–460

OECD (2010) Interim report of the green growth strategy: implementing our commitment for a sustainable future. Meeting of the OECD Council at Ministerial Level. OECD, Paris

Pedden M (2006) Analysis: economic impacts of wind applications in rural communities, 18 June 2004–31 Jan 2005. Subcontract report NREL/SR-500- 39099, contract no. DE-AC36-99-GO10337, prepared under subcontract no. LEE-4-44834-01. National Renewable Energy Laboratory, Golden

Ragwitz M, Schade W, Breitschopf B, Walz R, Helfrich N, Rathmann M (2009) The impact of renewable energy policy on economic growth and employment. The European Commission, Brussels

Reategui S, Tegen S (2008) Economic development impacts of Colorado's first 1000 megawatts of wind energy. National renewable energy laboratory. Conference paper no. NREL/CP-500-43505. Contract no. DE-AC36-99-GO10337. National Renewable Energy Laboratory, Golden

Roland-Holst D (2008) Energy efficiency, innovation and job creation in California. CUDARE working paper no. 1069. University of California-Berkeley, Department of Agricultural and Resource Economics. http://escholarship.org/uc/item/7qz3b977. Accessed 24 March 2013

Rutovitz J (2010) South African energy sector jobs to 2030. Prepared for Greenpeace Africa by the Institute for Sustainable Futures. University of Technology, Sydney

Sastresa EL, Uson AA, Bribian IZ, Scarpellini S (2010) Local impact of renewables on employment: assessment methodology and case study. Renew Sust Energ Rev 14:679–690

Scott MJ, Roop JM, Schultz RW, Anderson DM, Cort KA (2008) The impact of DOE building technology energy efficiency programs on U.S. employment, income, and investment. Energ Econ 30:2283–2301

Simas M, Pacca S (2011) Windpower contribution to sustainable development in Brazil. World Renewable Energy Congress, Linköping

Simons G, Peterson T (2001) California renewable technology market and benefits assessment. Technical report no. 1001193. California Energy Commission/Electric Power Research Institute, Sacramento/Palo Alto

Singh V, Fehrs J (2001) The work that goes into renewable energy. REPP research report no. 13. Renewable Energy Policy Project, Washington

Slattery MC, Lantz E, Johnson BL (2011) State and local economic impacts from wind energy projects: Texas case study. Energ Policy 39:7930–7940

Tourkolias C, Mirasgedis S (2011) Quantification and monetization of employment benefits associated with renewable energy technologies in Greece. Renew Sust Energ Rev 15:2876–2886

Ulrich P, Distelkamp M, Ulrike L (2012) Employment effects of renewable energy expansion on a regional level—first results of a model-based approach for Germany. Sustainability 4:227–243

UNEP/International Labour Organization/International Organisation of Employers/International Trade Union Confederation (UNEP/ILO/IOE/ITUC) (2008) Green Jobs: towards a decent work in a sustainable, low-carbon world. UNEP/ILO/IOE/ITUC, Geneva

Upadhyay H, Pahuja N (2010) Low carbon employment potential in India: a climate of opportunities. Discussion paper: TERI/GCN-2010:1. Center for Global Environment Research, The Energy and Resources Institute, New Delhi

Walsh J, White S (2008) Greener pathways: job and workforce development in the clean energy economy. University of Wisconsin-Madison, Center on Wisconsin Strategy, Madison

Wei M, Patadia S, Kammen DM (2010) Putting renewables and energy efficiency to work: how many jobs can the clean energy industry generate in the US? Energ Policy 38:919–931

Wicke B, Smeets E, Tabeau A, Hilbert J, Faaij A (2009) Macroeconomic impacts of bioenergy production on surplus agricultural land—a case study of Argentina. Renew Sust Energ Rev 13:2463–2473

Williams SK, Acker T, Goldberg M, Greve M (2008) Estimating the economic benefits of wind energy projects using Monte Carlo simulation with economic input/output analysis. Wind Energy 11:397–414

World Bank (n.d.) Issues in estimating the employment generated by energy sector 39 activities. Background paper for the World Bank Group energy sector strategy. World Bank, Washington

Ziegelmann A, Mohr M, Unger H (2000) Net employment effects of an extension of renewable-energy systems in the Federal Republic of Germany. Appl Energ 65:329–338

Chapter 6
Case Study Approach

Abstract In this chapter, the discussion shifts from EBED's foundations, definition, process, policies, and evaluations toward a practice-based setting to offer the reader a more detailed account of EBED through examples. Chapter 6 sets the stage for the case studies described in the following three chapters. This chapter reviews the criteria used to select the case studies and presents the range of case locations, approaches, energy types, economic development strategies, and policies.

We now transition from the first part of the book, which focused on the EBED domain, process, policy environment, and evaluation techniques, to the second half of the book, which focuses on EBED application. The purpose of the case studies described in Chaps. 7 through 9 is to extend the EBED framework discussion into a practice-based setting and thus to gain a deeper understanding of EBED through examples. Chapter 6 through 9 are particularly oriented toward the practitioner who is eager to understand how EBED works on the ground.

The case studies we present demonstrate how the EBED framework is applied in a variety of contexts, the manner in which programs incorporate energy and economic development goals, and how policies are used to support EBED efforts. The cases that we present cover a range of locations, approaches, energy types, economic development strategies, and policies.

The cases are divided into three chapters. In Chap. 7, we discuss subnational and local-level EBED projects. These "bottom-up" EBED projects exemplify how programs and projects are organized by regional, state, or local governments; nonprofits; entrepreneurs; and communities to meet EBED goals. We then summarize nationally driven or "top-down" EBED approaches in Chap. 8 to demonstrate how country-level strategies are shaped and executed. We review all of these examples with short snapshots in which we describe the program and apply it to the EBED framework described in the first half of the book.

In Chap. 9, we offer a more in-depth review of a hybrid "top-down" and "bottom-up" EBED strategy based on the ARRA. We provide more detail about the program, its funding, how it operated, and findings to date from Recovery Act efforts than in the previous two case study chapters. Although this chapter conveys

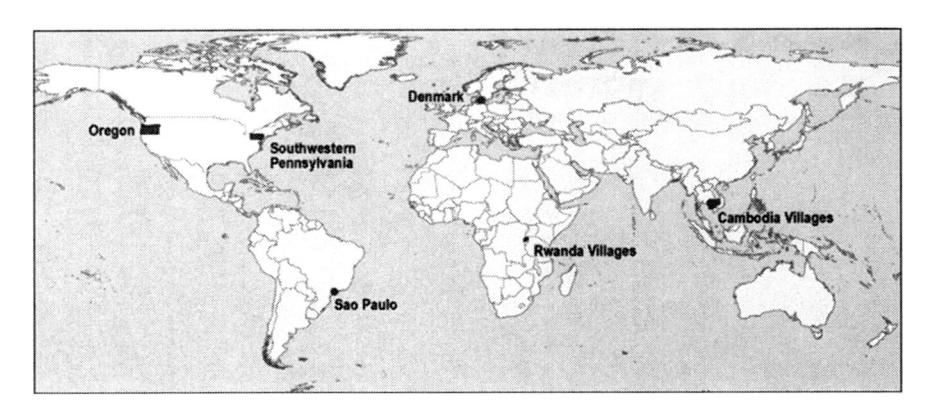

Fig. 6.1 Map of subnational case studies

the depth of one program's operations in this domain, the three chapters combined convey the breadth of EBED projects. In Chap. 10, we offer insights and conclusions from the cases and other considerations for advancing the EBED framework.

6.1 Selection of EBED Cases

We used four key criteria for selecting the cases. First, we considered geography, and aimed to ensure a diverse representation of global regions. We also selected both urban and rural locations. Of course, these programs are simply a sample of EBED; many countries and cultures are not covered in these cases.

Subnational cases represent the following regions: São Paulo, Brazil; Oregon, United States; Copenhagen, Denmark; rural villages in Cambodia; Southwestern Pennsylvania, United States; and villages in Rwanda. These regions are shown in Fig. 6.1.

National case studies are represented in Fig. 6.2 and include Singapore, China, Ethiopia, France, Laos, Morocco, and South Africa.

Our second criterion was that the cases fit the definition of energy-based economic development, that is, to have both energy and economic development goals. Although seemingly intuitive for inclusion in the EBED framework, we found many programs that claimed to meet dual goals, but on further examination fall short of actually doing so. For example, an effort by a country to increase energy efficiency in its building stock is an example of a federally driven strategy to reduce energy costs by increasing energy efficiency. However, if the program does not include any of the following, for instance, we did not consider it an EBED initiative:

6.1 Selection of EBED Cases

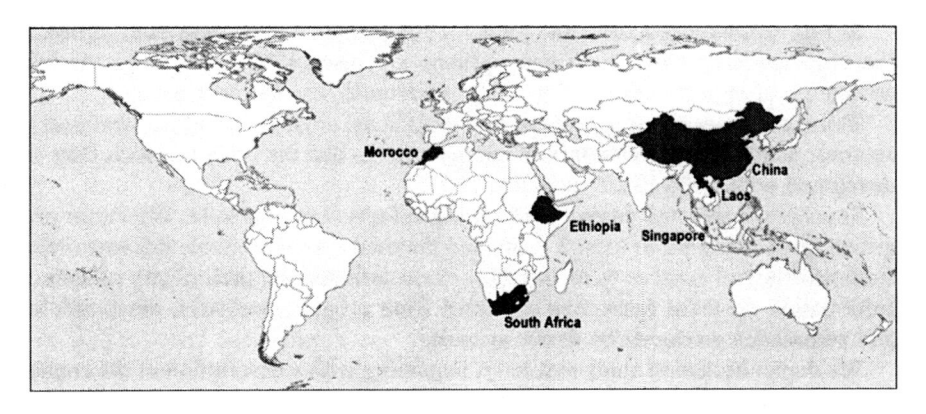

Fig. 6.2 Map of national case studies

Fig. 6.3 EBED goals by case

- workforce development to help construction workers learn energy efficient construction methods,
- innovation strategies to advance research, development, or manufacturing for low-emissions building products,
- export promotion strategies to market products or techniques for wider adoption,
- poverty reduction efforts to target low-income residents,
- supply chain development to expand local markets.

In Fig. 6.3, we summarize the joint EBED goals that each case exemplifies. In the corresponding case study descriptions, we discuss how the programs take advantage of opportunities at the energy-economic development nexus.

Third, the cases cover a range of low-emissions, advanced, and efficient energy sources, which illustrate the mix of energy options and the scale at which they are developed within the EBED domain.

Finally, we selected cases based on the information available. We chose programs that existed for at least 2 years and that had enough reliable documentation from a variety of sources to record their characteristics and preliminary outcomes. Information on these cases was extracted from program websites, news articles, and summaries produced by donor agencies.

We depict each case study similarly, beginning with a description of the context in which the program is implemented to demonstrate the need for EBED. We then describe the program in as much detail as the documentation allows, including timing; funding, actors, and institutions involved; and current status. We also highlight distinctive elements within each program and link program elements to EBED goals, approach, process, and trajectory. We underline supportive policies for the project and any potential implications from the project's history that may be important to note. We do not assess or evaluate these case studies for effectiveness. Although certainly important and an area for future exploration, this effort falls outside of the scope of this book.

Chapter 7
Subnational EBED Cases

Abstract Local or subnational EBED projects exemplify how programs and projects are organized to meet EBED goals by regional, state, or local governments; nonprofits; entrepreneurs; and communities. Local approaches are important because they can be initiated at a lower overall cost than national approaches, and they can be shaped to meet the specific needs of a community or region within a country. Further, when national governments are confined or restricted within their EBED strategy or agenda, subnational regions and localities can often break through those barriers and be laboratories of innovation for EBED. Subnational cases presented in this chapter are initiatives set in São Paulo, Brazil; Oregon, United States; Copenhagen, Denmark; rural villages in Cambodia; southwestern Pennsylvania, United States; and villages in Rwanda.

We advance the framework for EBED in this chapter with a review of locally or subnationally driven projects, programs, or initiatives. With the more localized focus, these initiatives tend to be smaller in scale in terms of both funding and potential effects. As discussed in Chap. 3, local approaches are important because they can be initiated at a lower overall cost than national approaches and can also be shaped to meet the specific needs of a community or region within a country. Further, when national governments are confined or restricted within their EBED strategy or agenda, subnational regions and localities can often break through those barriers and be laboratories of innovative practice. If necessary, locally oriented projects can serve as test beds for policies, programs, or initiatives that have the potential to be scaled later when national-level decision-makers are in a better position to implement them. Additionally, national-level programs that are not designed to meet local conditions and needs may be less likely to succeed than programs that are tailored specifically to local circumstances. For these reasons, we dedicate a chapter to an overview of subnational EBED projects.

S. Carley and S. Lawrence, *Energy-Based Economic Development*,
DOI: 10.1007/978-1-4471-6341-1_7, © Springer-Verlag London 2014

7.1 Case Study 1: The Bandeirantes Landfill Gas to Energy Project

Referred to as "South America's Economic Heart" (Forero 2012), São Paulo is also South America's largest city. The city has undergone an economic boom since macro-economic reforms were put in place in the 1980 and 1990s. Today it experiences the benefits of a diversified economy that is globally competitive and recently has attracted financial intellectual professionals from the United States and Europe (Forero 2012). However, this growth has brought significant stress to the city's infrastructure, and, some would argue, a widening gulf between rich and poor (Carroll 2008). Transportation and other infrastructure problems present such complications that elite businessmen and the mayor travel around the city by helicopter (Carroll 2008; Sharma 2011).

In addition to the city's infrastructure not keeping pace with growth and the burgeoning inequalities between rich and poor, urbanization has negatively affected the environment and local energy supply. As with any city undergoing rapid urbanization, these environmental and energy-related issues become intertwined. As the city continues to grow, stresses on energy supply and the environment create cyclical and complex problems for some residents and municipal service providers. For example, increased pollution in certain neighborhoods can affect the health and livelihoods of those residents and make it harder for them to maintain employment and education. To begin to address one of these stressors on local infrastructure—solid waste management—a public–private partnership turned one of the largest landfills in São Paulo into a waste-to-energy project to reduce GHGs, create economic opportunity and social development, and generate a local energy supply.

7.1.1 The Program

The Bandeirantes Landfill Gas to Energy Project (BLFGE) was created in 2004. In response to substantial CO_2 emissions from the landfill and subsequent health and environmental concerns, the City of São Paulo and Biogás Energia Ambiental S/A, a private company, created a partnership with the goal of capturing the methane gas emitted from the landfill and turn it into a biogas source of electricity provision. The goals of this project are to reduce the emissions of methane gas into the environment and to generate a source of revenue for the municipality. In turn, these goals will improve the environmental health for residents and produce an added source of municipal income. The income derived from the project can be used to invest in local infrastructure and programs to further advance positive social and environmental conditions, as well as economic opportunity for some of São Paulo's residents.

7.1 Case Study 1: The Bandeirantes Landfill Gas to Energy Project

The project has taken advantage of the Clean Development Mechanism of the Kyoto Protocol; the mechanism allows for the earned Certified Emission Reductions (CERs) by a country that can be counted as contributing to meeting Kyoto protocol goals to be sold for monetary value. Biogás Energia Ambiental S/A assumed the project's costs and business risks. It invested R$12 million for the gas capture system and secured R$45 million from a Brazilian Bank, Unibanco, for power plant construction. The city and Biogás split the project's profits in half. In 2007, São Paulo received 13,096,890 Euros as a result of selling their CERs to a Dutch bank. The municipality uses these funds to invest in a range of community programs that support a healthier environment, education, and other community development projects in this poor neighborhood. For example, apartment buildings for low-income residents of São Paolo were partly funded by BLFGE CER profits (Schmidt 2011).

As of 2009, the BLFGE is the largest landfill gas-to-energy project in the world, generating 170,000 MWh of energy each year. The energy from the project serves 400,000 city residents. The business strategy is to continue to turn waste into energy until 2018, when it is anticipated that the gas levels will not be adequate to continue to produce energy. In more recent years, Biogás also has replicated this project in São Joao (the eastern part of São Paulo) and in Rio de Janeiro (International Council for Local Environmental Initiatives (ICLEI) 2009).

7.1.2 EBED Framework

This project is noteworthy because of its comprehensive socioeconomic and environmental goals and aims. This case also demonstrates a public–private partnership, with the business approach largely driven by the private sector. Its success in one of the world's largest cities also provides ample opportunity to glean best practices and lessons learned for other rapidly urbanizing areas with solid waste management issues. It is uncertain, however, if or how other cities can leverage the CERs generated by the Kyoto Protocol to their benefit to have a similar return on investment as BLFGE.

This project's two stated goals overlap directly with the EBED goals to reduce GHG on the energy side and generate income for the municipality on the economic development side. The waste-to-energy project supplies efficient energy to 400,000 residents, reduces GHG emissions, improves local pollution levels, and invests accrued municipal revenues back into the local community. ICLEI (2009) reports that 26 jobs were created as a result of BLFGE's operations. Although certainly not a major employer for the neighborhood, these job outcomes, in conjunction with the resources invested to improve the community as a result of this project, bring economic benefits to this neighborhood near the Bandeirantes Landfill.

The project is tailored to improve this local community near the Bandeirantes Landfill and, as a result, has a place-based orientation. BLFGE is designed specifically to benefit the targeted neighborhood around the landfill through job creation and secondary benefits of GHG reduction. In terms of its scale of transformation, the project is linking to transformative: it links economic development benefits to an existing endeavor and it is transformative because it introduces waste-to-energy markets to this community. The most prominent policy mechanisms that support the project are the environmental climate change policies, as initiated through the Kyoto Protocol's Clean Development Mechanism and the use of CERs under this program.

This project demonstrates the potential of private–public partnerships to have a positive effect on energy, economic, social, and environmental needs of major urban centers. The ability to trade the CERs as a result of this project significantly heightens the economic benefits to both the city and the utility company. Cities that can take advantage of the CERs will find this project more beneficial to replicate than other cities that cannot reap the benefits from these "trades." This project has a 14-year time horizon and offers a significant source of GHG reductions over this time frame while providing a local source of energy.

7.2 Case Study 2: Clean Energy Works Oregon

The economic downturn of 2007–2009 affected a significant number of industries and regions across the United States. The construction industry was an industry that was heavily affected during this time. Prior to the economic recession, the construction industry was fueled by a steady demand from homebuyers and consequently supplied a substantial number of jobs with competitive salaries. However, economic activity ground to a halt with the mortgage crisis and subsequent shrinking of disposable incomes for potential homebuyers. In Portland, Oregon, for example, unemployment in the construction industry lingered around 30 % during this time period compared with the metropolitan area's annual unemployment rate across all jobs of 6 % in 2008, 10.7 % in 2009, and 10.5 % in 2010 (van der Voo 2010, revised 2011; U.S. Bureau of Labor Statistics, Local Area Unemployment Statistics n.d.). At the same time, the United States was engaged in two wars in Iraq and Afghanistan, respectively, which led to a growing awareness about the potential pitfalls of energy dependence and the need for a reduction in energy demand.

As U.S. legislators and the President developed a national economic stimulus plan, Portland, Oregon, offered an ideal launching pad for a bold energy efficiency program. Portland is a city that generally has a more progressive culture; accepts new ideas and practices more readily than other parts of the country; and, because of its location near the Cascade Mountains and Pacific Coast, also boasts a culture of environmental awareness and preservation.

Clean Energy Works Portland (CEWP), established in 2009, served as a pilot program that evolved into Clean Energy Works Oregon (CEWO). CEWP was a revolving loan fund program funded with $20 million in federal stimulus money from the ARRA. Its purpose was to use this funding to provide low-interest, long-term loans that helped residents retrofit their homes with energy efficient products and design. CEWP aimed to reduce energy consumption in participants' homes and facilitate job creation in the Portland area.

7.2.1 The Program

Based on the success of CEWP in the Portland area, the program was quickly scaled into CEWO to expand the same services to the state level. CEWO is operated by a nonprofit entity with an objective to retrofit homes for energy efficiency and create jobs in the construction industry by offering skills training in energy efficient installations. CEWO was launched with $18 million of the original funding from ARRA to CEWP to scale the program to 17 communities across the state. Over a 3-year period, CEWO plans to retrofit 6,000 homes and 3.5 million square feet of commercial buildings, create or retain 1,300 jobs, generate 300,000 MMBTUs in energy savings, and reduce measured life-cycle carbon dioxide by 200,000 metric tons (CEWO website 2012). To date, CEWO (2012) reports that it has completed close to 2,000 home energy projects and created 200 jobs in the construction industry.

The program also includes targets to incorporate under represented businesses and vulnerable populations as suppliers and workers for CEWO, in addition to ensuring quality pay and benefits as part of the program's elements (CEWO 2012). Public–private and nonprofit partnerships are integral to CEWO. Participating partners of CEWO include utility companies, local lenders, local governments, the Energy Trust of Oregon, the Oregon Department of Energy, and the U.S. Department of Energy.

7.2.2 EBED Framework

This EBED project is exemplary because it links intimately the goal of energy efficiency with job creation and retention. With its social inclusion targets for workforce training and small business suppliers, this case demonstrates an EBED initiative that not only focuses on energy and development outcomes, but also targets underrepresented populations in its program design. Finally, this model has demonstrated its ability to be scaled from a city level to the state level by reaching a variety of small- to medium-sized communities across a state.

CEWO embodies the EBED goals of energy efficiency and job creation and retention. It also targets poorer populations and underrepresented businesses. As a by-product of its energy reduction targets, CEWO has the potential to reduce carbon emissions in Oregon. The project represents a place- and household-oriented EBED approach that is linking in nature because it targets energy efficiency improvement for households as well as employment opportunities across the state of Oregon. It is somewhat transformative because it aims to expand the construction industry into an energy efficiency improvement niche industry. A combination of financing tools, in the form of low-cost loans and rebates with workforce training, comprise the supportive policy environment for CEWO.

To date, CEWO is tracking its impact and comparing outcomes to stated goals. This level of monitoring, evaluation, and subsequent communication provides important information to the program's investors and partnering financial institutions about the program's efficacy and return on investment. As the program evolves, it will be important to track CEWO activities and note whether the program can sustain its activities without federal government stimulus funds in the future. It will also be important to note if and how other regions can replicate this model without significant government investments.

CEWO has demonstrated noticeable potential for scale-up. It will be interesting to observe how other regions tailor this approach to their circumstances, such as regions with less public–private collaboration or in regions where the federal government is less active and local philanthropists and donor agencies play stronger roles in EBED.

7.3 Case Study 3: Copenhagen Cleantech Cluster

Denmark has few natural resources to supply its energy needs, which makes the country vulnerable to external prices and supply conditions. As explained by Moss (2012), the need to diversify its energy supply and make the country more energy efficient became evident over 40 years ago, during the oil shocks in the 1970s: "With few home-grown sources of energy, the country saw prices rocket, and this led to widespread acceptance that a new path needed to be taken ... Nuclear power was never seriously considered ... so long before other countries, Denmark began developing renewables."

Turning this commitment to building a renewable energy industry into an opportunity for economic growth, Danish universities and the government created the Copenhagen Cleantech Cluster (CCC). The purpose of the CCC is to seed and attract the world's leading clean technology business ventures. Coupled with the long-term national goal to make Denmark entirely reliant on renewable energy for its electricity, heating, industry, and transportation by 2050, Copenhagen, the country's capital city, is in the prime location to cultivate a global concentration of leading-edge clean technology industries (Danish Energy Agency 2012).

7.3.1 The Program

The CCC, initiated in 2009 (CCC 2013a), is a public–private organization comprising companies, research institutes, and government leaders. These participants are committed to creating an environment for relevant companies, researchers, and innovators to succeed in growing competitive ventures within the "cleantech" industry. According to Pernille Dagø, a research assistant with the CCC, the program was launched with total funding of 142 million DKK from the Capital Region of Denmark, Region Zealand, and the European Union Structural and Cohesion Funds, with an objective to make the CCC one of the most successful cleantech clusters in Europe.

The CCC has explicit goals and objectives. At the broadest level, the cluster aims to link cleantech technologies to community needs to foster broad-based industrial growth (CCC 2012). It aims to grow a comprehensive cleantech industry through the support of start-ups and the attraction of foreign companies to relocate in Copenhagen. More specifically, the Cluster's objectives from August 2009 to August 2014 are to create 1,000 jobs, attract foreign companies, seed 30 collaborative R&D endeavors, establish 15 international collaborations, create a 200 member-wide self-sustaining organization, support the growth of 25 entrepreneurs, and host at least 200 networking events (CCC website 2013b).

Dagø also relayed in e-mail communication that, by 2009, the cleantech sector was the fastest growing sector in the country with 720 companies, 120,000 employees, and revenues of over 40 billion Euros. For Eastern Demark, 522 companies were identified in 2010, which grew to 610 in 2011 and 750 in 2012. The CCC reports that 78 % of total employment growth from 2005 to 2011 for Eastern Denmark is attributed to the cleantech sector. Cleantech exports, however, have been negatively affected by the financial crisis, but there was a recent uptick in 2012 with expectations of continued positive trends in the years ahead.

Copenhagen's cluster development strategy is exemplary for its sheer scale, comprehensive scope, and a 40-year history of unified political and civic support for development and adoption of cleantech products. From an economic development perspective, the CCC has established a range of industry support organizations, including a national research lab, a research park, a business accelerator, and a gap financing mechanism, all of which are embedded in a broader triple-helix collaborative network of university, business, and government organizations. This model stresses efficiency with a "one-stop shop" approach to meet requests by cleantech businesses. The CCC also takes a proactive, outward-oriented approach that focuses on global market demand and international networks as it attracts industry and grows businesses within Copenhagen. For example, the CCC led the efforts to create the International Cleantech Network—an exclusive network of the world's leading cleantech clusters that share ideas to enhance business opportunities and competitive advantages across global regions (International Cleantech Network 2012).

7.3.2 EBED Framework

The CCC initiative emphasizes the economic aspects of EBED with a focus on increasing innovation and entrepreneurship, industry growth in cleantech sectors, and income growth resulting from the emergence of a competitive industry presence in the city. These economic goals coincide with energy goals to diversify the energy supply and bolster energy security. Relying solely on renewables generates an environmental benefit from enhancing a low-emissions environment.

The Cluster is mostly transformative in its trajectory by focusing its agenda on advancing growth in the clean energy sector and has leap-frog aspects on a technological innovation level. CCC is a result of the Danish government's innovation policies that support lowering the costs and risks of innovation, entrepreneurship, and technology commercialization. These kinds of policies tend to be foundational for successful innovation hubs such as the CCC. Cleantech is a priority target area of growth for the country too, and a variety of other national policies also support business and entrepreneurs that operate in the cleantech realm. Such policies include tax incentives for "green" innovation for cleantech products, financial support and grants for cleantech companies, and access to testing and demonstration facilities, among others.

Denmark's long-standing commitment to become entirely reliant on renewable energy has created an economic development climate that is remarkably supportive of seeding and growing clean technology markets in Copenhagen. Although significant barriers for the power companies and grid system operators to absorb and distribute renewable energy on a country-wide scale remain (Moss 2012), the political will exemplified by the Danish government to build its renewable energy industry and profit off of the transition is a major driver of change. Although the Cluster's unique political environment will be difficult for other regions to replicate, the CCC holds promise for allowing other regions to learn how to support niche industries relevant to their geopolitical circumstances and in support of EBED.

7.4 Case Study 4: Kamworks, Rural Cambodia

Similar to many other developing countries, Cambodia's energy needs are significant. Eighty percent of Cambodia's energy consumption comes from biomass, mostly in the form of timber, and only 20 % of the population can access the national power grid (Marks 2009). In addition, roughly 35 % of the population is poor and the poverty rate is as high as 46 % in rural areas (The Ministry of Industry Mines and Energy 2007). According to the Cambodian National Institute for Statistics (2008), 80 % of the country's population lives in rural areas, and just 13 % of these households have access to electricity as a main light source. Rural households, especially those that rely on agriculture for income, account for more than 91 % of the poor (The Ministry of Industry Mines and Energy 2007).

As Cambodia grows from a low-income to a middle-income country, it must improve access to sustainable and reliable energy. This goal is outlined in the country's Millennium Development Goals, established by the United Nations to meet the needs of the world's poorest, and in the country's National Strategy Development Plan, a blueprint to guide the country's social and economic growth (Marks 2009; The Ministry of Industry Mines and Energy 2007).

7.4.1 The Program

Kamworks is one of the leading solar energy companies in Cambodia, which strives to provide sustainable photovoltaic (PV) solar systems to households and businesses in Cambodia and other markets. The small business increases energy access in rural villages and manufactures its goods in-country (Kamworks 2012). Kamworks designs, installs, and maintains a range of solar products, such as off-grid systems for water pumps, electricity, and refrigeration, and grid-connected solutions for electricity and electrical back-up. The company also designs and manufactures products such as plug-and-play solar home systems and the Moon-Light solar lantern.

In all of its business endeavors, the company uses participatory methods to ensure that end users in the Cambodian market are the focus of their solar product development and dissemination. For example, before and during design phases of its product development, Kamworks visits local rural villages to conduct meet with villagers and inquire about the villagers' behaviors and willingness to use the products that Kamworks is designing. The villages are informed about the benefits of using renewable sources for power and become early consumers for product development and distribution. This participatory approach ensures Kamworks is demand oriented and simultaneously prepares its customers for new products they may not be already familiar with.

Established in 2006, Kamworks is a private social enterprise company that receives funding from its customers—individuals, households, and other kinds of organizations that purchase their products—to finance the company's efforts to scale solar solutions for individuals' electricity needs. The company received funding for two substantial projects in 2010 and 2011 from international donors that invest in the distribution of renewable lighting products to rural villages. This investment helped Kamworks expand its markets. Donors included the World Bank, the Energy and the Environment Partnership of the Mekong (EEP), the Ministry for Foreign Affairs of Finland, and the Nordic Development Fund (Kamworks 2012; EEP Mekong 2013).

These two larger-scale programs exemplify the potential impacts that social enterprises such as Kamworks can have on energy access and economic development. The first program, a 2-year pilot project funded with €136,997, is the

Kamworks lantern rental and solar home system financing program, which aims to introduce these technologies to rural Cambodian villages through rental schemes. The initiative Improving Access to Solar for Rural Electrification in Cambodia by Removing Rural and Technical Barriers Program is supported by EEP with the Cambodian Mutual Savings and Credit Network and Pico Sol as partners.

Kamworks expects the program to rent 3,000 lanterns to 100 village entrepreneurs to improve residential lighting at night. The company also expects to install 200 solar home electric systems through the rental scheme. In addition to engaging 100 village entrepreneurs, there will be five trained installers and 20 staff trained in solar loan finance (EEP Mekong 2013). As of January 2011, the program identified 75 rural entrepreneurs to pursue this micro-enterprise approach of solar lantern rental to villagers. To reduce the high costs of solar technology adoption, the program connects solar home customers with micro-financing options to make the initial capital investments for the products more affordable. In the second year of the program, Kamworks will focus on expanding the type of solar products it distributes through the rental program to additional villages throughout the country.

A second large initiative funded in collaboration with government and international donor support is the Cambodian Rural Electrification Fund. The Royal Government of Cambodia initiated this program with the support of the World Bank. The rural electrification fund supports the government's targets that "by the year 2020, all the villages throughout the Kingdom of Cambodia will have electricity of any form and by the year 2030, at least 70 % of all households will have access to grid quality electricity" (Cambodian Rural Electricity Fund n.d.). As part of this initiative, Kamworks will install 12,000 solar home systems across seven provinces in Cambodia. With financing schemes supported by the World Bank, Kamworks provided electricity in some of the most rural areas of Cambodia (Kamworks 2012).

7.4.2 EBED Framework

Kamworks is an example of a social enterprise approach to EBED in that it uses micro-finance methods to help customers afford solar homes. It uses a private-sector approach to improve energy access and generate economic benefits for Cambodians. Its focus on market need and the use of international donor support to scale up its products are distinctive paths for enterprises to pursue when connecting the goals and objectives of EBED to households and businesses in low-income regions.

Kamworks' efforts to produce and disseminate solar solutions in rural areas of Cambodia that lack access to energy mostly emphasize the EBED goals of reducing energy poverty and diversifying energy sources. However, the company displays an awareness of how to connect clean and efficient energy provision to entrepreneurship initiatives and the improvement of the company's market share.

7.4 Case Study 4: Kamworks, Rural Cambodia

For these reasons, Kamworks also fits the EBED goals of innovation and entrepreneurship and, to some extent, industry growth.

This case study exemplifies an initiative that embodies all three kinds of EBED approaches. Kamworks is market focused in that it is a private company with objectives to grow its market share in solar products. The initiative takes a household-focused approach because its market is targeted to those living in dwellings without energy access. Finally, the company takes a place-based approach because, through the Cambodian electrification fund, the primary distribution location of the lanterns is the villages in rural Cambodia. In terms of its EBED trajectory, Kamworks is transformative to leap-frog: it expands product lines into low- carbon sectors to provide entirely new products to new markets.

Four policies support a business model such as Kamworks. First, micro-lending policies enable village-level entrepreneurs to earn money by selling solar lanterns and homeowners to acquire the products with affordable loans. Second, tax and finance schemes provided by the Government of Cambodia's Rural Electrification Fund have supported the increased production and distribution of the company's products throughout rural Cambodia. Third, Kamworks engages in adoption efforts, although not necessarily policies, when they work with rural villages to provide information and tailor their products to these communities' needs. Finally, with the production, finance, and other business needs of a growing company, Kamworks has some workforce training elements, supported by donors and the government.

Market-based approaches that disseminate low-carbon, advanced, and efficient energy solutions and aim to reduce energy poverty are promising EBED strategies. Without international donors, however, social enterprises such as Kamworks may struggle to create and disseminate products at a price point affordable to the rural poor. A 2009 *New York Times* blog post about small-scale subsidies in Cambodia reports that "Generous subsidies for business and tax incentives for consumers are needed if developing countries like Cambodia are to promote renewable energy alternatives-particularly in rural areas" (Marks 2009). Large-scale hydropower and coal projects can secure financing more easily, but unlike Vietnam there are few, if any, tax incentives or loan subsidies to encourage renewable energy companies to enter the market in Cambodia (Marks 2009).

7.5 Case Study 5: Natural Gas Development in Southwestern Pennsylvania, United States

Since the 1980s, Pittsburgh, Pennsylvania, a U.S. city, has undergone massive economic restructuring. Pittsburgh was formerly heavily reliant on the steel industry, but lower production costs in other countries resulted in significant job losses due to the decline of steel and associated industries. Employment in durable goods fell by 30.8 % from 2000 to 2010, and jobs in steel and related industries fell 43.8 % from 1990 to 2010 (BLS 2013). Suburban and rural areas around Pittsburgh were especially hard hit because manufacturing jobs were more concentrated in these areas, whereas the city centers have been able to transform their economies with an influx of technology-based and professional service jobs. Like many other U.S. cities in formerly manufacturing-dominated states, the surrounding rural areas recovered more slowly, if at all.

The Marcellus Shale is a large rock formation in the Northeastern United States that contains large quantities of extractable natural gas. U.S. interest in natural gas exploration has surged in recent years as a result of three simultaneous trends: increased oil prices, a national call for greater energy security, and improved techniques for natural gas extraction. Natural gas also has been upheld by many in the energy industry as a reliable, efficient, and domestically supplied source of energy. Development of the Marcellus Shale for natural gas production offers Pittsburgh—and other similar regions—an opportunity to diversify its economic base, create new jobs, and generate investment to revive the local and state tax base. For places such as southwestern Pennsylvania, this economic opportunity coincides with the recent global recession, as well as a 20-year recovery from deeper economic shifts.

The economic and energy potential of natural gas extraction, however, comes with considerable controversy. Methods to extract the gas include mile-deep underground drilling and hydraulic fracturing, or "fracking." There is extensive debate about whether such methods are safe for the environment, public health, and the general quality of life for communities in which hydraulic fracturing occurs. Some argue that it is technically feasible to extract natural gas without harming the environment, while others believe the contamination from the chemicals used in hydraulic fracturing and the potential environmental degradation from drilling pose great risks. These issues cause much debate among stakeholders, as well as outside observers, and make this case one of the most controversial cases reviewed in this book. We fully recognize the contentious nature of this issue, and the inherent tension that this case presents between environmental protection and economic development. However, because our primary objective is to describe the EBED domain and provide examples of EBED in action, we do not explore the dimensions of this controversy in greater detail. We do recognize that natural gas production is an important topic within the EBED realm, and we have highlighted this case deliberately to demonstrate the full range of low-emissions, efficient, and advanced energy source case studies.

7.5.1 The Program

Natural gas industry development in southwestern Pennsylvania, since approximately 2007, has been driven by the private sector and supported by the government. As of 2013, 151 operators were drilling wells in the southwest region (Pennsylvania Department of Environmental Protection, Wells Drilled by Operator

7.5 Case Study 5: Natural Gas Development

Report 2013). Some of the more active natural gas companies in this region are Atlas Resources; CNX Gas Company; Range Resources Appalachia, LLC; and XTO Energy. Collectively, these four companies drilled 3,149 wells out of the 7,010 wells in the region between 2007 and 2013. Some of these companies, such as Range Resources, Chevron's Atlas, and ExxonMobil's XTO, are also major stakeholders in the U.S.-wide shale gas and oil industry.

Because of the potential to capitalize on natural gas activity for greater economic development, state and local governments have been working to harness natural gas's growth potential for state and local economic benefit. Simultaneous to natural gas business efforts, political leaders have crafted workforce development programs and tax and regulatory polices to help facilitate this industry's growth. For example, the Central Pennsylvania Institute of Science and Technology received a $2 million grant under Pennsylvania's Economic Growth Initiative to fund a facility for training natural gas and transportation workers (Shirk 2013). According to the Pennsylvania Department of Community and Economic Development (2013), businesses can receive training and consulting from advanced energy manufacturing centers. The state Department of Labor and Industry also provides labor and economic analyses of the Marcellus Shale industry, as well as profiles of careers in hydraulic fracturing through the Pennsylvania Statewide Career Coach website (Center for Workforce Information and Analysis 2012; Pennsylvania Department of Labor and Industry 2013). Moreover, as of 2012, Pennsylvania had one of the lowest drilling taxes in the country (The Pennsylvania Budget and Policy Center 2012).

Pennsylvania has a long history in the oil and gas industry, but natural gas production has increased significantly since 2009. Natural gas annual supply and disposition in Pennsylvania increased from 181,418 million cubic feet of dry natural gas production[1] in 2007–272,574 million cubic feet in 2009. In 2011, the year with most recent data, production was 1,301,661 million cubic feet (U.S. EIA 2013). The period December 2011 to August 2012 saw a 27 % increase in wells drilled and in natural gas production in the state (Pennsylvania Department of Community and Economic Development 2013).

Regional development of the Marcellus Shale has the potential to address acute economic development needs in rural Pennsylvania. Analysts and local community representatives estimate that drilling activity added $11.2 billion to regional GDP in 2010, with projections of $20 billion in 2020 (Considine et al. 2011). The direct economic impacts of constructing one well, including the costs of acquisition and leasing, permitting, site preparation, drilling, fracturing, completion, and production to gathering, are $7.6 million (Hefley and Seydor 2011). The prospect of land value increases also has local community members interested because they are likely to gain more financially as their land becomes more valuable. Further, the

---

[1] According to the U.S. EIA, dry natural gas production is the process of producing consumer-grade natural gas. It equals marketed production minus extraction loss. More information on definitions is at: http://www.eia.gov/dnav/ng/TblDefs/ng_sum_snd_tbldef2.asp.

state Department of Community and Economic Development (2013) reports that natural gas can benefit manufacturing and other important industries in the state because the by-products from natural gas production help in the manufacturing of medical, automotive, recreation, and construction products.

7.5.2 EBED Framework

Natural gas development from the Marcellus shale is a distinctive EBED strategy for two reasons. First, compared with states with long economic histories in oil and gas extraction—such as Oklahoma, Texas, North Dakota, and Wyoming—the state of Pennsylvania was not as prepared for the surge and scale of interest in natural gas extraction. Consequently, the state's workforce, infrastructure, and regulatory processes are being developed rapidly to meet current industry demands. For example, specialized training programs are required for workers within the regions of Marcellus Shale development so they can take advantage of the emerging job opportunities. Second, development of the Marcellus Shale highlights the pressures within the United States to diversify the country's energy base and provide low-cost, reliable energy resources.

From an economic development perspective, development of the Marcellus Shale for natural gas extraction increases household income as well as local revenues through taxes and fees, gross state product, and gross national product. Marcellus Shale development promotes industry growth as well. From an energy development perspective, shale gas diversifies energy sources and increases energy security. At a local level, many argue that shale gas is also a means for job creation. In contrast, opponents often cite the potential negative social impacts from a large temporary workforce and frequent use of heavy equipment in communities near the extraction sites. Some also purport that quality of life will decrease, while others counter that the investment from developing natural gas can help revitalize small rural downtowns.

Natural gas extraction is market and place based. Natural gas companies aim to expand their production and markets, and the state of Pennsylvania plans to diversify its economic base with an energy source that is in growing demand. This case study also exemplifies a linking to transformative strategy. As the state works through regulations to support the industry, an important component of this effort is the promotion of the economic development aspects of natural gas development. The case is somewhat transformative as well because shale gas development expands the state's energy economy into new energy markets.

A number of policies support the development of the natural gas industry at a subnational level. However, these policies do not conform to a single framework or originate from one executing agency. The state develops most policies on oil and gas drilling. The Pennsylvania Department of Environmental Protection (2013) states that oil and gas exploration and drilling fall under all or part of the state's oil and gas laws, the Clean Streams Law, the Dam Safety and

Encroachments Act, the Solid Waste Management Act, the Water Resources Planning Act, and the Worker and Community Right to Know Act. Strategies and policies related to economic development have been created and implemented by a network of local governments, industry representatives, and citizens. Federal, state, and local governments all participate in the development of the Marcellus gas industry and its regulations. Along with a U.S. EPA study of the environmental implications of expanded gas drilling, the federal Department of Labor extended grants to Pennsylvania, allocated by state agencies and nongovernmental organizations, for workforce development (U.S. EPA 2011; ShaleNet n.d.). Local governments and the state also receive new revenues from "impact fees" which they then reallocate to the development of local infrastructure such as roads, schools, and hospitals and emergency management (State Impact 2011).

Development of the Marcellus Shale is a means to both diversify the regional energy base through expansion of the natural gas industry and improve the regional economy through job and local wealth creation. Pennsylvania leads the way in development of the Marcellus Shale for EBED. Other states such as New York, North Carolina, and Ohio continue to watch closely to learn from Pennsylvania's successes and failures as they consider a similar development path. The ability of the industry and public leaders to respond to environmental and human health concerns associated with hydraulic fracturing may prove crucial to the EBED potential of natural gas development.

7.6 Case Study 6: Nuru Energy

Energy poverty affects over two billion people worldwide (Business United Nations n.d.). The impacts from the lack of energy are acute in rural areas of developing countries because of the lack of basic energy alternatives needed for food preparation, clean water, and light, for example. The need for low-cost solutions that address energy poverty is urgent, because economic and social development cannot advance without access to energy. The need for cleaner energy is also critical. Households often use kerosene for lighting, which results in respiratory illness and contributes to local air pollution and general carbon dioxide emissions. Kerosene is also expensive. In Rwanda for example, households spend an average of one-third of their incomes on the fuel (Africa Enterprise Challenge Fund 2012; Nuru Energy 2012).

7.6.1 The Program

Nuru Energy is a for-profit social enterprise, "built on a belief in hyper-local economic communities and in environmental preservation through economic development and technological innovation" (Singh 2011, p. 1). Nuru Energy developed a portable off-grid modular lighting system and recharging platform called the POWERCycle™, a human-powered pedal generator. The rechargable lights and power generators produced by Nuru Energy can replace kerosene fuel and lanterns (Kolodny 2010). Economic development aspects of the program arise through Nuru Energy's distribution system, a recruited network of village-level entrepreneurs who buy 100–200 lights and a POWERCycle to start their own recharging stations. These microenterprises make marginal profits off the lights and recharges, which produces both jobs and income in poor, rural regions.

Started with a seed grant from the World Bank, Nuru Energy attracted $500,000 in grants to fund its 2.5-year start-up phase in pilot communities in Rwanda. With several international prizes from the United Nations in 2010 and commercial financing from Bank of America Merrill Lynch in 2011, Nuru Energy expanded operations into East Africa and India, with offices in Kenya, Uganda, and India as of 2013 (Nuru Energy 2012).

The Urwego Opportunity Bank helps advance the microenterprise model in Rwanda. The bank provides an asset loan to each entrepreneur as a partner so that he or she can purchase lights and a POWERCycle pedal generator. The entrepreneur has 6 months to pay back the loan (The Rwanda Focus 2010). According to Samer Hajee, the CEO of Nuru Energy, entrepreneurs "[make] $1.50 for 20 min of charging. That's what they made earlier by working the whole day" (Singh 2011, p. 2). Moreover, customers only pay for the light and 27 cents for a charge that lasts for 40 h, compared with $1.75 for 1 l of kerosene, which lasts 13 h (Africa Enterprise Challenge Fund 2012). Nuru Energy estimates an overall cost savings of 85 % for households that switch from kerosene to the rechargeable lights. Those individuals that run the Nuru Energy micro-franchises also make on average over $3 USD per day, which is three to four times the national daily average wage in Rwanda (Millington 2011).

After the 10-month trial in Rwanda, Nuru Energy sold over 500 lights, and ten micro-franchise owners operated with a profit (Millington 2011). In 2010 Nuru distributed over 10,000 lights and expanded to work with 70 village-level entrepreneurs (Nuru Energy 2012). As of 2011, over 5,000 households in Rwanda and India adopted the Nuru Energy lighting model (Millington 2011). Nuru Energy plans to have 10,000 village-level entrepreneurs participate in its business model in order to distribute 1.8 million lights by the end of 2016 (Nuru Energy 2012).

7.6.2 EBED Framework

Nuru Energy uses a private-sector business model that fits all of the EBED dimensions and aims to positively affect energy poverty, economic development, education outcomes, and health conditions for households. Nuru Energy's business model aims to reduce energy poverty through increased access to energy for energy-poor households. It targets rural and remote villages in developing countries by providing localized, reliable energy, with resulting cost-savings and job opportunities. Because Nuru seeks to displace the household use of kerosene with human-powered lighting systems, Nuru also fits the EBED goal of energy diversification. Through reduction or elimination of kerosene use in households, the likely by-product of this EBED program is improved health for women and children, who tend to stay in the household during the day. According to a quote by Sameer Hajee in *Forbes*, "Our product helps reduce the use of kerosene, a significant cause of respiratory diseases. We're helping the local environment by removing the fumes and toxicity of kerosene from the air. We are creating job opportunities for the community. Plus for the first time the kids in the community now have the ability complete schoolwork at their leisure, freeing up for time for play and extracurricular!" (Singh 2011).

Nuru's business model is to introduce and disseminate a new product, its lighting system, in the rural village marketplace. Thus, this social enterprise model fosters entrepreneurship and innovation. The program has an explicit goal of job creation because it targets locals as employees. The program reduces poverty because poor households benefit from significantly lower energy costs. The promise of Nuru lies in the fact that its business leaders study the markets they try to reach to make sure their product design meets local needs and leverages local market opportunities. Nuru Energy collects regular data on households served, products sold, and micro-franchises.

This case falls within the household- and market-oriented EBED approaches because Nuru addresses energy poverty and household incomes, but innovation and entrepreneurship drive its business model. The innovative nature of the program also places Nuru Energy on a leap-frog trajectory. The program takes a new and minimally tested innovation and introduces it into new markets. Microenterprise and micro-franchise policies support Nuru and enable the entrepreneurs to invest in the power cycle and other capital and start-up costs. As a result of this investment, entrepreneurs can distribute and sell the lighting systems and chargers in rural villages.

Four years after the development of a prototype and tests in Rwandan villages, Nuru is expanding to other parts of Africa and India. With the adoption of the lights and chargers in each new location, Nuru's management works to understand and mitigate the unique market barriers presented in each location. For example, Mr. Hajee reports that the process of market penetration in India faced more obstacles than in Rwanda. Because fewer institutions provide microfinance in India, implementation of the micro-franchise model for distribution was more difficult.

References

Africa Enterprise Challenge Fund (2012) Nuru Energy Rwanda. React Focus, October 23. http://www.aecfafrica.org/newsletter/REACT_Focus_Newsletter_23_October_2012.pdf. Accessed 8 Nov 2012

Business United Nations (n.d.) Expanding rural energy entrepreneurship and access to clean lighting in East Africa. http://business.un.org/en/commitments/302. Accessed 11 Nov 2012

Cambodian National Institute of Statistics (2008) General population census of Cambodia 2008. http://www.nis.gov.kh/nis/census2008/Census.pdf. Accessed 6 Jan 2013

Cambodian Rural Electricity Fund (n.d.) http://www.ref.gov.kh/eng/index.php?fn=InformationE.htm. Accessed 5 Jan 2013

Carroll R (2008) City of the future. The Guardian, March 13. http://www.guardian.co.uk/news/2008/mar/14/rorycarroll.insidebrazil2. Accessed 30 May 2013

Center for Workforce Information and Analysis (2012) Marcellus shale occupational compendium. http://www.pacareercoach.pa.gov/PA_MarShale_Occ_Compendium.pdf. Accessed 1 Aug 2013

Clean Energy Works Oregon (CEWO) (2012) Clean Energy Works high road outcomes: new faces, career pathways and increasing influence. http://www.cleanenergyworksoregon.org/wp-content/uploads/2012/09/HighRoad_Short_090612.pdf. Accessed 9 May 2013

Considine TJ, Watson R, Blumsack S (2011) The Pennsylvania Marcellus natural gas industry: status, economic impacts and future potential. Pennsylvania State University College of Earth and Mineral Sciences, Department of Energy and Mineral Engineering, University Park

Copenhagen Cleantech Cluster (2012) The global cleantech report 2012. Copenhagen Cleantech Cluster, Copenhagen

Copenhagen Cleantech Cluster (2013a) About Copenhagen Cleantech Cluster. http://www.cphcleantech.com/home/about. Accessed 9 July 2013

Copenhagen Cleantech Cluster (2013b) Q & A about Cleantech. http://www.cphcleantech.com/about/q–a-about-cleantech. Accessed 9 July 2013

Danish Energy Agency (2012) Danish climate and energy policy. http://www.ens.dk/en/policy/danish-climate-energy-policy. Accessed 10 Nov 2012

Energy and Environment Partnership with the Mekong Region (2013) Energy and environment partnership with the Mekong region. http://eepmekong.org/index.php?reload. Accessed 5 Jan 2013

Forero J (2012) Expats lured by Brazil's booming economy. The Washington Post, February 5. http://www.washingtonpost.com/world/americas/expats-lured-by-brazils-booming-economy/2012/01/13/gIQA4jnasQ_story.html

Hefley WE, Seydor SM (2011) The economic impact of the value chain of a Marcellus shale well. Working paper, Katz Graduate School of Business, University of Pittsburgh Joseph M. Katz Graduate School of Business, Pittsburgh. http://www.business.pitt.edu/faculty/papers/PittMarcellusShaleEconomics2011.pdf. Accessed 27 Sept 2011

International Cleantech Network (2012) http://www.internationalcleantechnetwork.com/. Accessed 10 Nov 2012

International Council for Local Environmental Initiatives (ICLEI) (2009) Turning pollution into profit: the Bandeirantes landfill gas project. ICLEI case studies 2009 #107, Sao Paulo, Brazil. http://local-renewables.org/fileadmin/sites/local-renewables/files/04_Local_Practice/01_Case_studies_and_Descriptions/Sao%20Paulo%20107%20High%20Res.pdf

Kamworks (2012) http://www.kamworks.com. Accessed 5 Jan 2013

Kolodny L (2010) Start-ups win with plans to displace disposables. New York Times, April 13. http://boss.blogs.nytimes.com/2010/04/13/start-ups-win-competition-with-plans-to-displace-disposables/. Accessed 8 Nov 2012

Marks, S (2009) In Cambodia, a cry for small-scale subsidies. New York Times, November 2. http://www.green.blogs.nytimes.com/2009/11/02/in-cambodia-a-cry-for-small-scale-subsidies/?_r=0. Accessed 5 Jan 2013

Millington A (2011) The opportunity project. Winner spotlight: Nuru Energy. Our impact, December 9. http://www.theopportunityproject.org/blog/story/Nuru%20Energy. Accessed 11 Nov 2012

Moss P (2012) Are Denmark's renewable energy goals wishful thinking? BBC News Science and Environment, April 8. http://www.bbc.co.uk/news/science-environment-17628146. Accessed 10 Nov 2012

References

Nuru Energy (2012) History. http://nuruenergy.com/about-nuru/our-vision-mission. Accessed 8 Nov 2012

Pennsylvania Department of Community and Economic Development (2013) Marcellus and Utica shale fact sheet. http://www.newpa.com/webfm_send/3057. Accessed 2 May 2013

Pennsylvania Department of Environmental Protection (2013) Wells drilled by operator report. http://www.depreportingservices.state.pa.us/ReportServer/Pages/ReportViewer.aspx?/Oil_Gas/Wells_Drilled_By_Operator. Accessed 30 April 2013

Pennsylvania Department of Labor and Industry (2013) Marcellus shale fast Facts. http://www.portal.state.pa.us/portal/server.pt?open=514&objID=1222103&mode=2. Accessed 1 Aug 2013

Schmidt KUB (2011) Trading methane for housing. Development in a changing climate, June 17. http://blogs.worldbank.org/climatechange/trading-methane-housing. Accessed 9 July 2013

ShaleNet (n.d.) About ShaleNet. http://www.shalenet.org/About/Index. Accessed 15 Dec 2011

Sharma R (2011) Brazil, the Un-China. Time, July 18. http://www.time.com/time/magazine/article/0,9171,2081831,00.html. Accessed 9 June 2013

Shirk E (2013) Governor Corbett Announces Expansion of Central Pennsylvania Institute of Science and Technology Training Facility; Creates 185 Jobs. Pennsylvania Department of Community and Economic Development. http://www.newpa.com/newsroom/governor-corbett-announces-expansion-central-pennsylvania-institute-science-and-technology-training. Accessed 1 Aug 2013

Singh A (2011) How Sameer Hajee has shed real light in Africa. Forbes, October 10. http://www.forbes.com/sites/csr/2011/10/10/how-sameer-hajee-has-shed-real-light-in-africa/. Accessed 8 Nov 2012

State Impact (2011) What the new impact fee law, Act 13, means for Pennsylvania. NPR State Impact, PA, December 14. http://stateimpact.npr.org/pennsylvania/tag/impact-fee/. Accessed 14 Dec 2011

The Ministry of Industry Mines and Energy (2007) Residential energy demand in rural areas. http://www.kamworks.com/uploads/tx_news/2007_10_09_MIME_Residential_Energy_Demand_in_Rural_Areas.pdf. Accessed 6 Jan 2013

The Pennsylvania Budget and Policy Center (2012) PA enacts among the lowest natural gas drilling fees in the nation. http://pennbpc.org/gas-drilling-tax. Accessed 1 Aug 2013

The Pennsylvania Department of Environmental Protection (2013) Laws, regulations and guidelines. http://www.portal.state.pa.us/portal/server.pt/community/laws%2C_regulations___guidelines/20306. Accessed 2 May 2013

The Rwanda Focus (2010) Bringing light to Rwanda: Nuru Energy wins prestigious environmental award. The Rwanda Focus, March 29

U.S. Bureau of Labor Statistics (BLS) (2013) Databases, tables, & calculators by subject: employment. http://www.bls.gov/data/#employment. Accessed 31 July 2013

U.S. Bureau of Labor Statistics. Local Area Unemployment Statistics (n.d.) Portland-Vancouver-Hillsboro, OR-WA Metropolitan Statistical Area. http://data.bls.gov/pdq/SurveyOutputServlet;jsessionid=07131D13C2D0A01C9D86904B9E031EF1.tc_instance4. Accessed 20 July 2013

U.S. Energy Information Administration (2013) Natural Gas annual supply and disposition by state. http://www.eia.gov/dnav/ng/ng_sum_snd_a_EPG0_FPD_Mmcf_a.htm. Accessed 2 May 2013

U.S. Environmental Protection Agency (2011) EPA's study of hydraulic fracturing and its potential impact on drinking water resources. http://www.epa.gov/hfstudy/index.html. Accessed 15 Dec 2011

van der Voo L (2010) Clean Energy Works expands into 2011. Sustainable Business Oregon, December 22. http://www.sustainablebusinessoregon.com/articles/2010/12/clean-energy-works-expands-into-2011.html?page=all. Accessed 6 April 2013

Chapter 8
National Case Studies

Abstract National EBED initiatives are much larger in scale in terms of funding and intended reach than the subnational efforts described in the previous chapter. Many of these top-down strategies are driven by economic motivations such as the reduction of energy costs for a country; improved energy access; or enhanced growth in a low-emissions, efficient, and advanced energy sector. Most of these programs are driven by national governments in contrast to some of the entrepreneurial-driven programs at the subnational level. This chapter presents cases from Singapore, China, Ethiopia, Laos, Morocco, and South Africa.

As a complement to the subnational cases reviewed in Chap. 7 this chapter reviews nationally driven EBED initiatives. All of these initiatives are large in scale in terms of funding allocated and their intended reach. Many of these top-down strategies are driven by economic motivations: to reduce energy costs for a country, improve energy access, or target growth in a low-emissions, efficient, and advanced energy sector, or all three. Most of these programs are driven by national governments, in contrast to some of the entrepreneurial-driven programs covered in Chap. 7.

8.1 Case Study 7: Biofuels in Singapore

Singapore is a leading innovative economic engine in Southeast Asia. Between the 1960 and 1980s, the city-state led the way, along with Hong Kong, South Korea, and Taiwan, to become one of the "newly industrialized countries." Singapore accomplished its economic transformation within a very small geographic area through an export-oriented economy, a well-educated workforce, a strong physical infrastructure such as ports and roads, and a business-friendly investment environment (EconomyWatch 2010). Government officials also blended a heavily market-oriented approach with significant government intervention in the economy to guide the growth of new and globally competitive industries in sectors such as manufacturing and finance (MTI Insights 2011a). In fact, this market- and government-based approach to development was so exemplary that it was coined 'the Singapore Model' (EconomyWatch 2010).

One of the prominent elements in the Singapore model is investment in new, innovation-based initiatives. The Singapore model began an economic tradition in the country of pioneering new initiatives that embraced outside markets and new ideas for investment in emerging competitive sectors. As part of these innovation-led growth strategies, Singaporean government officials invested significantly in new biofuels refining technology and production facilities to establish the country as a regional leader in biofuels R&D. These investments were intended to also create a hub for biofuel refinement that nearby countries such as Indonesia, Thailand, Myanmar, Malaysia, and the Philippines could take advantage of The Economist (2006). The initiative successfully established Singapore as a regional leader in biofuels refinement and as a country that embraced low-emissions fuels and renewable energy. In addition to the visionary economic development aspects of Singapore's biofuels initiative, the investments in biofuels is also consistent with Singapore's energy policy framework, which aims to support low-emissions technologies and industries (MTI Insights 2011b).

8.1.1 The Program

The "NexBTL" foreign direct investment (FDI) project was a national public–private investment initiative to develop a "next-generation" biofuels refinery plant in Singapore. The plant cost S$973 million and started operation in November 2010 with a capacity of 800,000 metric tons, making it the largest biodiesel plant in the world at the time. Neste Oil, the pioneering Finnish R&D renewable diesel production firm, led the project (Yahya 2011). As of late 2012, the leading German airline, Lufthansa, was testing the biofuels from the refinery for its jets, which indicates commercial viability of the biorefinery plants (The Star 2012).

The NexBTL plant is notable for its size, its reliance on R&D and innovation, and the risk that the public and private investors placed on biofuels as an emerging and significant globally competitive industry. This EBED initiative puts Singapore's biofuels industry on a potential trajectory to be a global leader in next-generation biofuels technology. The NexBTL project is part of a larger plan by the Government of Singapore to invest in low-carbon energy as a targeted growth sector. In 2007, government officials announced a target of S$1.7 billion in clean energy-sector development, along with 7,000 jobs by 2015. Biofuels are one of four areas targeted to capitalize on the opportunities in the clean energy sector and help sustainable energy providers expand in the region (MTI 2007). The investment in biofuels is in line with the EBED goals of industry growth, national income expansion through FDI, and innovation. Biofuels also diversify the country's energy sources (MTI 2011b).

8.1.2 EBED Framework

The Singaporean biofuel's effort has both a place-based and market-based orientation: it aims to increase growth in biofuels in Singapore by positioning the city-state as a strategic location for biofuels R&D and refineries. The project is transformative to leap-frogging in nature in that it both expands the country's growth into low-carbon markets and relies on untested technologies of biofuel refinement to create new products for the airline industry among other users of jet fuels. Singapore used policy tools similar to those that made the country the economic powerhouse it is today. This case involves tax, finance, and regulatory tools that facilitate FDI, R&D, and nascent industry growth, all of which foster a healthy environment for biofuels to compete in the global marketplace. The government invested S$350 million in 2007 in clean energy in general, and a Clean Energy Programme Office funds several programs; the country also relies on its existent oil industry to provide infrastructure and chemical product support for the biofuels industry (Singapore Economic Development Board 2013). In addition, at the opening of the NexBTL plant in 2011, Deputy Prime Minister and Minister for Defense, Teo Chee Hean, noted that the country's Economic Development Board was improving research institutions and creating public–private research partnerships (Hean 2011).

The country's investment in NexBTL and the biofuels industry more generally marks a major effort by Singapore to become a leader in the development of low-carbon fuels, and is in keeping with their global and national policy priorities to create an alternative energy sector, build new industries for domestic and export markets, and create new job opportunities. The initiative exemplifies all aspects of EBED; energy and economic development goals are interconnected with equal weight placed on both disciplines.

This effort also reveals some inherent tensions between the expansion of the country's economy and preservation of the environment. The NexBTL project, in particular, features a theme that is consistent with many biofuels operations, where there are trade-offs between environmental factors and low-carbon fuel development. The Singapore project attracted controversy, for example, because of its planned reliance on Malaysian and Indonesian palm oil as the primary feedstocks. Palm oil plantations are a leading cause of deforestation in Southeast Asia; deforestation contributes to GHG emissions, climate change, and other negative consequences as a result of environmental degradation. The refinery will increase demand for palm oil, because it will require 200,000 hectares of palm oil plantations to supply 1 million tons of oil per year. To counter this, Neste Oil committed to the exploration of alternative fuels to reduce the reliance on palm oil. For example, Neste pledged to put 80 % of its profits into R&D to develop new and alternative fuel stock sources. Waste animal fats imported from Australia and New Zealand are one example of alternative fuels that are being tested (Neste Oil 2008).

8.2 Case Study 8: China Golden Sun

The Chinese economy grew from a GDP of under $1 trillion in 1997 to more than $8 trillion in 2012. Despite recent news that China's manufacturing sector is slowing in growth, it remains evident that the sector has been one of the key drivers of economic transformation in the country's extraordinary growth trajectory over the past few decades (AFP 2013; Wassener 2013). Similar to other countries, China has also been looking for ways to mitigate some of the economic problems caused by the recent recession, while simultaneously fueling its continued economic growth.

The country's total energy consumption is the largest in history, but China's energy resource availability per capita is below the world average, even lower in some cases than other developing countries (Ma 2013). Government officials have to provide enough power to fuel an economy that grew from under $1 trillion in 1997 to more than $8 trillion in 2012. As of 2013, China imported 60 % of its oil and 30 % of its natural gas. Coal is the country's main supply of domestic energy (Ma 2013). China's need to support its manufacturing sector, address sharp declines in demand for manufactured products due to the recession, and diversify the country's energy resource base, made the "Golden Sun" program a timely initiative for government leaders to undertake.

8.2.1 The Program

China's "Golden Sun" program was a national-scale initiative designed to support and expand a nascent and struggling solar PV panel industry. The program's purpose was to block the closure of 10,000 domestic solar PV manufacturers during the recession when demand from Europe and Asia fell dramatically (Ying 2011). At the same time, the program aimed to stimulate the adoption of renewable energy in the country and spur domestic demand for PV panels. The program relied on large-scale subsidies to support demonstration projects for solar technologies. The goal of these demonstration projects was to accelerate technological progress of solar PV panels and introduce these technologies into broader markets (Ministry of Finance of the People's Republic of China, Program Announcement 2009).

The Golden Sun program provided upfront subsidies for qualified PV projects. Subsidies ranged from 50 % of the costs of purchase and installation to up to 70 % in remote regions where projects were not connected to the power grid (Ying 2011). In November 2009, government officials announced the provision of 20 billion Renminbi (RMB)—at the time $2.93 billion USD—of subsidies for 294 Golden Sun demonstration projects that totaled 642 MW of power. The subsidies were provided over a 2-year period, until 2011. The program targeted all regions in the country by holding application rounds for submissions within each province.

According to initial requirements, every province was allowed to apply for financial support for up to 20 MW of solar power, although this did not always occur in practice (Ying 2011). Companies awarded subsidies had to follow several requirements, including a 3-year time horizon to complete their projects; pricing requirements for silicon panels; efficiency requirements; guaranteed energy outputs after 2, 10, and 25 years; and a demonstration of success. Most projects were industrial or commercial in nature and had the potential for solar energy use on site. Eighteen projects were off-grid and 35 consisted of large power plants set to channel electricity straight to the grid (Wang 2009). Large firms such as LDK and Suntech used some of the subsidies, but small micro-enterprises also benefited from the subsidies (Wang 2009; Ying 2011).

8.2.2 EBED Framework

The Golden Sun program was distinctive for the magnitude of its targeted support for a low-emissions energy industry sector. Golden Sun exemplified an EBED effort created almost entirely for economic development reasons but inclusive of strong energy-related features. Golden Sun's design closely aligned with the 11th 5-Year Plan (2008), which set aggressive national renewable energy targets for the installation of 300 MW of solar energy by 2010 (National Development and Reform Commission 2008). The EBED case is an example of a broader and more comprehensive initiative, in which the government sought to meet economic development objectives through the support and expansion of the energy sector. In so doing, the expanded availability and affordability of solar energy options led to both businesses and households adopting renewable energy. This program also reportedly played a significant role in the support of 10,000 domestic solar PV businesses that were at risk of closing during the recession (Ying 2011).

Similarly to many of the national case studies, this approach was mainly market oriented and place based. Some elements targeted household benefits, where the solar power demonstration projects supported residential dwellings, but they were relatively small compared with the market focus of the overall initiative. The program was transformative in that it sought to create new low-emissions domestic markets for solar power installations.

Expansive subsidies comprised the key policy features of this industry growth program. More broadly, Golden Sun fit within China's 5-Year Economic Development Plan to create a new industry for China and expand the global market for solar energy while accomplishing national renewable energy goals. The effort to target small micro-enterprises also revealed supportive entrepreneurship policies (National Development and Reform Commission 2008).

After the 2-year program concluded in 2011 and many of the demonstration projects completed, Chinese leaders instituted a national feed-in-tariff program. The feed-in-tariff program targeted large-scale solar projects and set a 10 GW PV installation target for 2015, a 10-fold increase in solar capacity at the time. This policy was arguably an extension of Golden Sun to continue to help pave the way for more solar production and adoption in China (Davidson 2011; Liu 2011).

Now that the demonstration projects are mostly over, several questions have surfaced about the program's effectiveness. There are three reported concerns with the Golden Sun program. First, cost structure requirements had different impacts for companies of different sizes. Large companies reported that the cost requirements forced them to sustain losses over time from their demonstration projects, while small micro-enterprises referred to Golden Sun as a "lifeline" (Ying 2011). Second, beyond mere subsidization, the initiative revealed the importance of skill development for the design, installation, and maintenance of on-grid PV projects. This workforce capacity need was a constraint in the short run but was targeted as an area for training and job growth opportunities in the future (Davidson 2011; Liu 2011). Third, despite these massive public efforts, representatives from the solar industry reported that local domestic demand to support this industry sector's growth remained weak. It will be interesting to track if the large feed-in tariff will effectively spur greater domestic demand for solar power. The initiative also highlighted the need for stronger policy guidelines and standards to improve planning and oversight for renewable energy projects (Ying 2011).

It is important to note that international critics of Golden Sun and related programs identified this type of heavily subsidized industrial policy as providing unfair national subsidies for Chinese renewable energy manufacturers and enabling these firms to export their products at unrealistically low prices. Export competitiveness of Chinese PV was a by-product of the Golden Sun initiative, but the aims of the program were foremost to expand the renewable energy manufacturing sector and drive solar project development for domestic installation at a time when export markets were in decline.

8.3 Case Study 9: Ethiopia National Cookstoves Program

Ethiopia faces energy poverty and declining natural resource stocks of biomass fuels. The energy sector in Ethiopia is bifurcated into those with modern energy access (i.e., electricity and petroleum) and those with traditional energy access (i.e., biomass). A majority of the population falls into the latter classification. Eighty percent of Ethiopia's population lives in rural areas and performs small-scale agricultural activity. Most households in the country use firewood, charcoal, dung, and crop residues for household heating and cooking (EU Energy Initiative—Partnership Dialogue Facility 2013). Overall, 89 % of total energy consumed in the country was biomass as of 2010 (EU Energy Initiative—Partnership Dialogue Facility 2013). In urban areas, 94 % of total energy demand is met through biomass, and in rural areas, this percentage is about 99 % (Ethiopian Environmental Protection Agency 2004, in Beyene and Koch 2013). This significant consumption of biomass has not only led to excessive deforestation and contributed to global climate change, but it has also created difficulties in the agricultural sector. As demands for other forms of biomass increase for energy use, dung and other natural fertilizers in agriculture are rising in cost and becoming limited in supply. The result of these trends is a tension between the demand for biomass as energy and the demand for biomass as an agricultural commodity. The use of biomass for indoor cooking and heating also contributes to indoor air pollution, which creates deleterious health effects for those who are exposed to the smoke (Federal Democratic Republic of Ethiopia, Ministry of Water and Energy (MoWE) 2012).

In response to these vast challenges, Ethiopia embarked on a national "Climate Resilient Green Economy Strategy." A core element of this initiative is Ethiopia's national cookstoves program. The program sets goals to create domestic markets for cleaner cookstoves that use less biomass and create less indoor air pollution, in addition to spurring an industry for cookstove manufacturing and distribution through small micro-enterprise development. The program represents an EBED case study that leverages basic applications of technology to improve product development for small micro-enterprises and to create jobs.

8.3.1 The Program

MoWE is the lead agency that established a public- and private-sector initiative to support the national cookstoves program in partnership with the United Nations Development Program, the BARR Foundation, the Global Alliance for Clean Cookstoves, and others. The program takes a commercial approach to creating a market for cleaner cookstoves by supporting new stove manufacturers and supply chain development on the market side and funding education initiatives to cultivate Ethiopian households as customers. The education initiatives combine with consumer credit programs to mitigate the financial barriers to purchasing and adopting stoves.

The MoWE program sets goals for developing and disseminating at least 9 million improved cookstoves to more than 4.5 million households by 2015. It also aims to save 2.1 tons of woody biomass per year per household with a total abatement potential of 14 megatons of carbon dioxide equivalent. Program designers estimate that this will stop 1,000–2,000 deaths per year caused by indoor air pollution and will create 5,000 private-sector jobs (Federal Democratic Republic of Ethiopia, Environmental Protection Authority). The program is expected to increase rural household income by as much as 10 % with cost savings from using cleaner cookstoves (MoWE 2012).

8.3.2 EBED Framework

This project integrates multiple programmatic components that aim to deliver a range of socioeconomic, energy, and environmental benefits. The creation of small industrial cookstove manufacturers will generate jobs and stimulate entrepreneurship. Improved stove dissemination will also deliver household cost savings by reducing firewood use, which will result in environmental benefits as well. Carbon reductions will occur through a reduction in CO_2 emissions from biomass combustion. And health benefits will accrue from less indoor air pollution exposure for those who spend more time in the household, which generally is women and children.

This program address all points of intervention—households, markets, and place. In recognition of the country's dependence on biomass, the project creates markets for clean cookstoves for dissemination and use in Ethiopian households. The Ethiopia cookstoves program is transformative in nature in that it advances cleaner energy applications within existing sectors.

This case study primarily features financial policies. First, the Government of Ethiopia provides, or at least plans to provide, financial support of the cookstoves program from a combination of government budgetary allocations, concessionary loans and grants, funds from international and bilateral development agencies, and market-friendly lending policies from local commercial banks. The program also will pursue carbon financing options from the CDM and voluntary credit markets as improved cookstoves are used more widely (MoWE 2012; Accenture 2011).

The Ethiopian cookstoves project has potential to reduce energy poverty and GHG emissions and also create jobs and enhance entrepreneurship opportunities. Improved cookstove dissemination in Ethiopia in the last several years already demonstrates the ability to employ thousands of individuals, mostly women, with the creation of jobs and business opportunities for both the urban and rural poor (MoWE 2012). Distinct from some of the other cases, this program also affects women's and children's health by reducing indoor air pollution.

8.4 Case Study 10: Lao People's Democratic Republic National Hydropower Initiative

The small, mountainous country of Lao People's Democratic Republic (PDR) embraces energy development for industrial growth, national income, and poverty alleviation. One of the poorest countries in Southeast Asia, Lao PDR has sought avenues for economic growth since it liberalized its markets in the 1990s following the Soviet Union's collapse and the decline of its communist hold on Lao PDR. Less than 5 % of the land in Lao PDR is suitable for subsistence farming, because it is a landlocked country rich in natural resources and largely covered by forests. The farming sector, however, employs 80 % of Laotians. Most people outside Vientiane, the capital, do not have access to basic resources such water or electricity (BBC News Asia Pacific 2012).

The Lao Government, with support from the international community, targeted hydroelectric power generation as the thrust of the country's major national development strategy from the late 1980s onward (Molle et al. 2009). With assistance from neighboring countries such as Japan, China, and Vietnam and international development organizations such as the Asia Development Bank, the nation embarked on an economic effort to develop hydropower for export to Thailand and other neighboring countries. This effort was coupled with a national push for broader economic reforms in the country (BBC News Asia Pacific 2012). Over the past two decades, the country transformed itself into a lower-middle income economy by reducing poverty levels and increasing national income. From 2006 to 2012, the gross national income doubled, and the hydropower sector was crucial to this economic growth (ADB 2013).

8.4.1 The Program

The Lao hydropower program is a nationally driven agenda to develop renewable energy over the course of 100 years (Whaley et al. 2009). Laos has a hydropower potential of 26,000 MW, of which 10 % is currently exploited. The National Socio-Economic Development Plan of 2011–2015 calls for over 5,000 MW in additional hydropower with goals of electrification and electricity export. Lao PDR signed agreements to export at least a combined 8,500 MW of electricity to Thailand, as well as 10,000 MW to Vietnam (Liying 2012). The Nam Theun 2 (NT2) dam, a $1.3 billion USD 1070 MW dam supported by loans from the ADB and World Bank, is the flagship project in this national program (Ferrie 2010; BBC News Asia Pacific 2012).

The Lao PDR hydropower program illustrates a primary focus on energy as an economic development strategy. Lao PDR views hydropower as the principle economic engine for fueling growth and development. Hydropower production in Laos is planned as a means of acquiring income from exports to neighboring countries. Income growth from these exports can then be channeled toward national development initiatives such as poverty alleviation and other economic development-related investments. The case also illustrates the influence of international donors and international investors in steering Laos toward hydropower. Technical and financial assistance was provided through a range of bilateral and multilateral grants and loans to realize this development strategy.

8.4.2 EBED Framework

The national agenda to develop and promote hydropower fits into several energy and economic development goals. The most relevant EBED goals are to increase industry growth and increase national income through energy exports. The national initiative also aims to reduce energy poverty by expanding the country's energy infrastructure to provide energy outside of the capital city. The approach to this effort is mostly place based but also involves some market-based elements, given that it concentrates on the development of Lao PDR's hydropower infrastructure to sell electricity through competitive, market-driven approaches. This initiative is further characterized from an EBED perspective by its linking nature—it connects existing natural resource development to economic development opportunities.

Supportive policies have played an important role in the development of hydropower in Lao PDR. The National Electricity Law of 1997 and the Lao PDR Government reforms of key laws and regulations to facilitate greater private-sector participation in developing the country's vast hydropower resources were critical for hydropower development. In addition, with bilateral and multilateral donors, Lao PDR's national economic development policy included industry policies that focused on creating a business climate conducive for investment in and expansion of the energy sector.

The Nam Theun 2 project, and the broader hydropower project portfolio in Laos, seems to be a positive economic development engine for the country that has facilitated both rural electrification and funding for social development in rural areas in Laos. The hydropower exports are also contributing to industrial development in neighboring countries.

It is important to note, however, that this aggressive development of hydropower in Laos also brought considerable controversy. Human rights and environmental concerns often took second priority to the major dam projects. Critics note that environmental impacts to domestic flora and fauna or the social implications of forced relocation were not adequately considered in dam development projects. Dams in Laos displaced villagers from their homes and led to loss of agricultural, fishing, and forest areas (Ferrie 2010). Water resources specialists also note the impact these projects have on the hydrology of the Mekong Basin and its tributaries, and the impact on riparian countries and communities across the lower Mekong basin (The Economist 2012).

8.5 Case Study 11: Morocco Solar and Wind

Morocco depends on imports for 97 % of its energy needs, so it is highly dependent on other countries for energy supplies (Coats 2012). According to the Director of Morocco's Renewable Energy Agency, these economic conditions stress the country's balance of trade, which escalated significantly during the 2000s when the price of petroleum increased (Burger 2012). Similarly to other North African countries, especially on the heels of the Arab Spring, Morocco is also focused on job creation as a means to ensure greater employment for its growing youth and working age populations. To respond to some of these challenges, Morocco created a bold renewable energy strategy linked to industry and job creation efforts. With an abundance of solar and wind resources, its strategic geographic position in North Africa, and strong links to Southern Europe, Morocco targeted solar and wind energy development for domestic energy supply and export to Europe.

8.5.1 The Program

In 2009, Morocco's King stated that the country would institute solar and wind programs to meet 42 % of the country's demand for electricity from renewable energy sources by 2020. The announcement came with a job creation strategy that estimates 50,000 new jobs will be generated as part of the country's renewable energy growth plan (Burger 2012). Fourteen percent of the renewable energy is planned to come from solar sources, 14 % from wind, and 14 % from hydropower (Coats 2012). Organizational support, legal frameworks and other supportive strategies followed the announcement to foster the growth of solar and wind energy production.

The solar program launched with a $9 billion USD investment in a solar power project with targets of 2,000 MW in generating capacity, 1 million tons fewer oil imports per year, and 3.7 million tons fewer carbon dioxide emissions annually. The Moroccan Agency for Solar Energy (MASEN), a public–private venture, was established to lead the solar project's implementation. By the year 2020, the project aims to establish solar electricity production in five major sites: Ouarzazate, Ain Bni Mathar, Foum Al Oued, Boujdour, and Sebkhat Tah. Two technologies, concentrated solar power (CSP) and PV, will be used in these project sites (Kingdom of Morocco, Moroccan Investment Development Agency). A projected $1 billion solar power farm was launched with a $300 million investment from the World Bank and will generate 2 GW of new capacity (Burger 2012). The country's solar program also features a solar water heater initiative that targets the installation of 440,000 m^2 of thermal solar sensors in 2012 and 1.7 million m^2 in 2020. This program is projected to avoid 920,000 tons of CO_2 emissions per year and create hundreds of permanent jobs in the renewable energy sector (Kingdom of Morocco, Moroccan Investment Development Agency). Job creation strategies focus on small-scale distribution and infrastructure development and on strategic points within the green energy value chain (Ettaik 2012; Burger 2012).

The Moroccan Integrated Wind Energy Project will lead wind energy development. A total investment of $3.86 billion for a 10-year period will help the country increase its wind energy capacity from 280 MW in 2010 to 1,000 MW in 2015 and 2,000 MW in 2020, in addition to attracting $3.87 billion in investment (Kingdom of Morocco, Moroccan Investment Development Agency n.d.; Frontier Market Network 2013). A wind power equipment factory will manufacture parts such as turbines, blades, nacelles, and spare parts (Burger 2012). A recently announced $621.32 million wind farm will add 300 MW to the national grid and provide 40 % of the country's wind energy (Frontier Market Network 2013).

Workforce training elements for both solar and wind programs include specialized courses in electrical engineering at national schools and universities. New training for the next generation of energy technicians at Morocco's vocational training institutes and R&D skills building that align with industry demands are also planned. To enable the export-led growth of renewable energy to Europe a high-voltage direct current transmission line to Europe is under construction

(McGinniss 2012). Along with other countries in Northern Africa, the Desertec Industrial Initiative ("Dii") plans to link renewable power from sun-rich countries to European markets. Dii broke ground in 2012 on a 500 MW solar plant in the Moroccan desert, the first of many renewable energy power stations that could collectively supply 20 % of Europe's electricity by 2050 (Coats 2013; Wasserrab 2013). At the time of writing, however, several corporate backers exited this initiative, and it is unclear if this arrangement will be maintained over time (Coats 2013).

8.5.2 EBED Framework

This project is significant for its scale, the progressive policy environment put in place to enable its success, the large-scale investments made by Morocco, and the establishment of public–private partnerships with foreign companies and governments. It is also worth noting that the country's leadership did not deviate from this strategic priority during the events of the Arab Spring, when the demands on government resources were high. The Kingdom's pro-business orientation combined with regulatory reforms to promote domestic power production and electricity exports are key elements that enabled the initiative.

Morocco's renewable energy strategy contains multiple energy and economic development goals, so it represents one of the more comprehensively focused EBED projects. Morocco's strategy seeks to improve energy security and diversify energy sources, while also generating income through foreign direct investment and exports. Job creation goals are also important, as achieved through the training and employment programs for grid infrastructure. The programs are market and placed based, with some minimal targets to households. The programs are also transformative to linking in nature because they both expand renewable energy markets while creating entirely new export sectors for the country.

Policies that support the development of the renewables sector include the Energy Law, drafted in 2009 and finalized in 2010, which allowed the country to undertake the renewable energy initiative. The Law enabled competition for electrical energy production and allowed small- and medium-sized power projects to connect to the national grid. The Law also enabled public–private partnership development and created the ability to export renewable energy via the national electricity grid and its international interconnections to Algeria and Spain.

The Energy Law also introduced new and fundamental institutional changes for renewable energy development. MASEN was established as a result of a public–private partnership to ensure the execution of the Moroccan Solar Plan. The National Agency for the Development of Renewable Energy and Energy Efficiency created a policy arm dedicated to renewable energy development. AMISOLE, the Moroccan Industry Association for Solar Industries and Wind Energy Companies, was also created by the government to support the business interests of Moroccan industrialists in the sector and support workforce development

(Coats 2012; Ettaik 2012; Kingdom of Morocco, Moroccan Investment Development Agency n.d.).

A $1 billion USD donation from the Kingdom of Saudi Arabia, the United Arab Emirates, and the Hassan II fund for social and economic development created an energy development fund to help co-finance these projects. The Kingdom is considering using the CDM to support growth in this sector as well.

This project demonstrates the significance of policy and public–private collaboration in facilitating improvements in Morocco's business climate for renewable energy investment. Morocco's pro-business orientation also generated excitement among the international community, which attracted multilateral development partners and private-sector investors into the sector. Without this reform and positive enabling environment, it would not have been possible for Morocco to attract the needed capital to execute such massive solar and wind project investments. However, some international businesses seem to be pulling out of the export-related deals. It will be interesting to observe if this is a minor or significant event and to determine the causes for these fluctuations in the businesses' strategies over time (Coats 2013). Similar to the Copenhagen case study in the prior chapter, this case also points to an example where committed and consistent political will played a major role in an EBED program.

8.6 Case Study 12: South African Renewables Initiative

South Africa is a middle-income country with a diverse set of challenges and opportunities. The country is highly dependent on coal and petroleum to drive its energy- and carbon-intensive industries. South Africa emitted approximately 491 metric tons of CO_2 equivalent in 2005, the most recent year of data available at the time of writing (World Resources Institute 2013). In 2009, then-President Jacob Zuma announced that the country would reduce its carbon emissions to 34 % by 2020 and to 42 % by 2025 (South African Department of Energy 2011b). Despite South Africa's relatively high level of export earnings and industrial production, the country faces socioeconomic challenges such as general high unemployment, high youth unemployment, income inequality, and a high proportion of low-wage and low-quality jobs. The South African Department of Economic Development's 2010 New Growth Path Framework notes that the country still ranked among the top ten countries with the lowest level of employment in the world as of 2010 (Economic Development Department, Republic of South Africa 2010).

Beginning around the early 2000s, South African leaders recognized that the untapped potential for renewable energy in wind and solar resources was a viable opportunity to develop a comprehensive strategy to meet climate change agreements with an aggressive whole-of-government approach and to deliver a new low-carbon EBED strategy to address some of the persistent socioeconomic challenges faced by the country.

8.6.1 The Program

The South African Renewables Initiative (SARi) is a public-sector program launched in 2010 to guide South African industrial and energy strategy in solar and wind development. Spawned from an interministerial committee on energy and industry, SARi targets the creation of new domestic and export-ready renewable industries. To bolster the ability of the South African government to finance this significant endeavor, in 2011 an international partnership between the governments of South Africa, the United Kingdom, Germany, Denmark, and Norway and the European Investment Bank made an announcement to attract long-term funding to build out infrastructure and foster growth in the renewable industry (South African Department of Trade and Industry 2010). Since the program's announcement in 2010 and the creation of the investment partnership in 2011, the government commissioned prefeasibility and feasibility studies to ascertain the opportunities, costs, and obstacles to developing the renewable sector (South African Department of Trade and Industry 2011; Zadek et al. 2010).

SARi's programmatic goals are embedded within three government strategies: the Ministry of Economic Development's New Growth Path Framework, the 2011/2012–2013/2014 Industrial Policy Action Plan (IPAP 2), and the 2010 Integrated Resource Plan for Electricity. These plans suggest several intended goals for SARi. According to the economic development strategy, the country aims to generate 300,000 direct "green economy" jobs by 2020, including approximately 80,000 in the manufacturing sector and the remaining in construction, operations, and maintenance. The plan also aims to improve worker skills by adding at least 30,000 engineers by 2014 and increasing public- and private-sector R&D spending from 0.93 % in 2007 to 2008 to 1.5 % by 2014 and by 2 % by 2018 (Economic Development Department, Republic of South Africa 2010).

The IPAP 2 requires greater energy efficiency and use of renewable energy in buildings to help spark demand for solar water heaters, energy efficient lighting, and wind turbines, among other low-emissions and efficient products. The plan also emphasizes the need to increase manufacturing capabilities in the "green sector," including goals to stimulate demand, reduce electricity costs, and construct centralized solar power demonstration plants (South African Department of Trade and Industry 2011).

Within the nation's 2010 Integrated Resource Plan for Electricity, government officials requested adding approximately 19 GW of renewable energy capacity to the grid by 2030, mostly in wind, solar photovoltaics, and centralized solar power (South African Department of Energy 2011a; SARi 2011). In October 2010, investors and other stakeholders began discussing the possibility of a solar park in the Northern Cape. The park plans include both solar power generation plants and facilities for manufacturing its components. Estimated to cost 150 billion Rand, the World Bank committed 1.8 billion Rand to the project. At the time of writing, it is unclear if or to what extent the project is attracting additional financing. Assuming that it is funded, the initiative is estimated to generate 5,000 MW and would create 12,300 construction jobs annually over 10 years. Once constructed, the plant will create 3,010 operations and maintenance jobs (South Africa Info, 2010).

8.6.2 EBED Framework

The SARi strategy emphasizes the importance of environmental, cultural, and social preservation. The initiative also seeks to simultaneously develop the renewable energy sector and pursue "inclusive growth" from skill development and job creation for the chronically unemployed. This focus on "inclusive growth" is noteworthy, because it demonstrates the use of energy-based development to enhance a country's social and political stability.

This project illustrates a national-level case study of energy-based development designed to drive industry and job growth. SARi uses public-sector investment, along with the support of international donors, to attract private-sector interest and investments in the renewable energy industry. A proposed suite of policy measures will also support this effort.

Energy diversity, energy security, the reduction of GHG emissions, and industry development are SARi's main energy and economic development goals. Because these goals target markets for renewable energy in South Africa, the initiative is place and market oriented. SARi is transformative to leap-frogging in nature because it expands South African exports into renewable energy sectors, which are relatively new areas for the country. The policy environment to support SARi is robust. Bilateral agreements, regulations, systems management, national public funds, low-cost loans, a technology transfer fund, state-owned enterprises, and the CDM are employed to support South Africa's growth in renewable energy (South African Department of Trade and Industry and Department of Energy 2011).

SARi is an EBED initiative that demonstrates significant energy and development potential. At the time of writing, the program remains largely in the conceptual stages with relatively few reported projects underway, despite the fact that international financial commitments from other governments are in place (Mclaughlin 2012). However, as the South Africa Department of Trade and Industry and the Department of Energy (2011) note in their report about SARi, "Effective integration of renewables into the nation's broader policy framework, particularly with respect to economic development, is a challenge in itself." It will be interesting for EBED practitioners to follow how the SARi initiative develops over time and observe if and how it affects renewable energy development in the wider Sub-Saharan African region, along with the economy of the area.

References

Accenture Development Partners for the Global Alliance for Clean Cookstoves (2011) Enhancing markets for delivery of improved cookstove development and promotion support in Ethiopia: market analysis, recommendations and program plan. http://www.cleancookstoves.org/resources_files/ethiopia-market-assessment-report.pdf Accessed 6 July 2013

AFP (2013) China manufacturing sees 1st fall in 7 months: HSBC. The China post, May 24. http://www.chinapost.com.tw/china/national-news/2013/05/24/379421/China-manufacturing.htm. Accessed 19 June 2013

Asia Development Bank and Lao People's Democratic Republic fact sheet (2013) http://www.adb.org/sites/default/files/pub/2013/LAO.pdf. Accessed 19 June 2013

BBC News Asia Pacific (2012) Laos profile. BBC news Asia Pacific, 5 November. http://www.bbc.co.uk/news/world-asia-pacific-15351898. Accessed 19 June 2013

Beyene AD, Koch SF (2013) Clean fuel-saving technology adoption in urban Ethiopia. Energ Econ 36:605–613

Burger A (2012) Morocco stays renewable energy course amidst Arab spring. Clean Technica, July 18. http://cleantechnica.com/2012/07/18/morocco-stays-renewable-energy-course-amidst-arab-spring/. Accessed 3 July 2013

Coats C (2012) Is Morocco the Mediterranean's green energy savior? Forbes, April 18. http://www.forbes.com/sites/christophercoats/2012/04/18/is-morocco-the-mediterraneans-green-energy-savior/2/. Accessed 5 July 2013

Coats C (2013) Has the death of a Euro-African renewable link been exaggerated? Forbes, June 14. http://www.forbes.com/sites/christophercoats/2013/06/14/has-the-death-of-a-euro-african-renewable-link-been-exaggerated/. Accessed 5 July 2013

Davidson M (2011) China solar tariff a really big deal. National Resource Defense Council. In: Green Economy, August 9. http://uk.ibtimes.com/articles/20110809/china-solar-ariffreally-big-deal.htm. Accessed 22 June 2013

Economic Development Department, Republic of South Africa (2010) The new growth path framework. http://www.economic.gov.za/communications/51-publications/151-the-new-growth-path-framework. Accessed 1 July 2013

EconomyWatch Content (2010) Singapore economy. EconomyWatch, March 18. http://www.economywatch.com/world_economy/singapore/. Accessed 19 June 2013

Ettaik Z (2012) Renewable energy in Morocco: collecting data. Redaf Irena workshop. http://www.irena.org/documentdownloads/events/marrakechmay2012/2_zohra_ettaik.pdf. Accessed 6 July 2013

EU Energy Initiative–Partnership Dialogue Facility (2013) Biomass Energy Strategy (BEST) Ethiopia. http://www.euei-pdf.org/country-studies/biomass-energy-strategy-best-ethiopia. Accessed 26 Oct 2013

Federal Democratic Republic of Ethiopia, Environmental Protection Authority (2012) UN Conference on Sustainable Development (Rio +20) national report on Ethiopia. http://sustainabledevelopment.un.org/content/documents/973ethiopia.pdf. Accessed 30 June 2013

Federal Democratic Republic of Ethiopia, Ministry of Water and Energy (MoWE) (2012) Scaling-up renewable energy program Ethiopia investment plan. http://www.epa.gov.et/Lists/News/Attachments/17/SREP%20Ethiopia%20Investment%20Plan%20-%20Version%20for%20External%20Review%20-%20January%202012.pdf. Accessed 30 June 2013

Ferrie J (2010) Laos turns to hydropower to be "Asia's battery." The Christian Science Monitor, July 2. http://www.csmonitor.com/World/Asia-Pacific/2010/0702/Laos-turns-to-hydropower-to-be-Asia-s-battery. Accessed 19 June 2013

Frontier Market Network (2013) Morocco: renewable energy drive gains pace. Frontier Market Network, April 16. http://www.frontiermarketnetwork.com/article/1766-morocco-renewable-energy-drive-gains-pace#.UdPt0pxhRyV. Accessed 5 July 2013

References

Hean TC (2011) Speech by Deputy Prime Minister and Minister for Defence Mr Teo Chee Hean at Neste Oil Singapore NExBTL Renewable Diesel Plant Opening Ceremony. http://www.edb.gov.sg/content/dam/edb/en/news%20and%20events/News/2011/Downloads/Neste-Oil-Corporation-opens-NExBTL-biorenewable-diesel-plant-in-SG-Speech-by-Mr-Teo-Chee-Hean.pdf. Accessed 1 Aug 2013

Kingdom of Morocco, Moroccan Investment Development Agency (n.d.) Why Morocco: Morocco in brief. http://www.invest.gov.ma/?lang=en&Id=9. Accessed 5 July 2013

Liu C (2011) China uses feed-in tariff to build domestic solar market. New York Times, September 14. http://www.nytimes.com/cwire/2011/09/14/14climatewire-china-uses-feed-in-tariff-to-build-domestic-25559.html. Accessed 30 June 2013

Liying S (2012) Laos: the hydroelectric battery of Southeast Asia. ESI Bull 5:11

Ma D (2013) China's search for a new energy strategy. Foreign affairs, June 4. http://www.foreignaffairs.com/articles/139425/damien-ma/chinas-search-for-a-new-energy-strategy. Accessed 5 July 2013

McGinniss P (2012) Morocco leading the world towards a green energy future. EcoWatch, November 23. http://ecowatch.org/2012/morocco-green-energy-future/. Accessed 5 July 2013

Mclaughlin, P (2012) SARi Renewables initiative slow to gather momentum. Parly report, May 19. http://parlyreportsa.co.za/uncategorized/sari-renewables-initative-slow-to-gather-momentum/. Accessed 7 July 2013

Ministry of Finance of the People's Republic of China (2009) Notice on the implementation of the Golden Sun demonstration project (Golden Sun Program Announcement). http://jjs.mof.gov.cn/zhengwuxinxi/zhengcefagui/200911/t20091116_232594.html. Accessed 19 June 2013

Ministry of Trade and Industry Singapore (2011a) MTI insights 1965-1978. http://www.mti.gov.sg/MTIInsights/Pages/1965---1978.aspx. Accessed 19 June 2013

Ministry of Trade and Industry Singapore (2011b) MTI insights—energy: a changing energy landscape: the energy trilemma. http://www.mti.gov.sg/MTIInsights/Pages/Energy-.aspx. Accessed 19 June 2013

Ministry of Trade and Industry Singapore (MTI) (2007) Energy for growth. http://www.mti.gov.sg/ResearchRoom/Documents/app.mti.gov.sg/data/pages/885/doc/NEPR%202007.pdf. Accessed 30 June 2013

Molle F, Foran T, Floch P (2009) Introduction: changing waterscapes in the Mekong Region—historical background and context. In: Molle F, Foran T, Floch P (eds) Contested waterscapes in the Mekong Region: hydropower, livelihoods, and governance. Earthscan, London

National Development and Reform Commission (2008) Renewable energy development, Eleventh Five-Year Plan. http://www.sdpc.gov.cn/nyjt/nyzywx/W020080318390887398136.pdf. Accessed 1 July 2013

Neste Oil (2008) NExBTL renewable synthetic diesel. http://www.climatechange.ca.gov/events/2006-06-27+28_symposium/presentations/CalHodge_handout_NESTE_OIL.PDF. Accessed 19 June 2013

Singapore Economic Development Board (2013) Alternative energy. http://www.edb.gov.sg/content/edb/en/industries/industries/alternative-energy.html. Accessed 1 Aug 2013

South Africa Info (2010) SA woos investors in solar energy. BuaNews In: South Africa Info, October 29. http://www.southafrica.info/business/investing/opportunities/solarpark-291010.htm

South African Department of Energy (2011a) Integrated resource plan for electricity. http://www.doe-irp.co.za/content/IRP2010_promulgated.pdf. Accessed 30 June 2013

South African Department of Energy (2011b) The launch of the South African Renewables Initiative. UNFCCC COP 17 climate change conference, Durban. http://sarenewablesinitiative.files.wordpress.com/2011/12/sa-renewables-initiative-launch-7-december-2011.pdf. Accessed 30 June 2013

South African Department of Trade and Industry (2011) 2011/2012–2013/2014 Industrial Policy Action Plan (IPAP 2). http://www.info.gov.za/view/DownloadFileAction?id=144975. Accessed 30 June 2013

South African Department of Trade and Industry and Department of Energy (2011) Partnering for green growth: SARi update briefing. http://sarenewablesinitiative.files.wordpress.com/2011/

03/sari-brochure-parterning-for-green-growth-for-distribution-071211.pdf. Accessed 30 June 2013

The Economist (2006) Green power in Southeast Asia: fuels rush in. The Economist, August 24. http://www.economist.com/node/7833918. Accessed 19 June 2013

The Economist (2012) Damming the Mekong River: river elegy. The Economist, November 3. http://www.economist.com/news/asia/21565676-laos-admits-work-going-ahead-controversial-dam-river-elegy. Accessed 19 June 2013

The South African Renewables Initiative (SARi) (2011) Declaration of intent on the South African Renewables Initiative (SARi). http://sarenewablesinitiative.files.wordpress.com/2011/12/sari_declaration_of_intent_011211-as-printed.pdf. Accessed 30 June 2013

The Star (2012) Lufthansa looking at palm-based biofuel to reduce carbon footprint. In: Eco-Business, August 16. http://www.eco-business.com/news/lufthansa-looking-at-palm-based-biofuel-to-reduce-carbon-footprint/. Accessed 19 June 2013

Wang U (2009) Here comes China's $3B, "Gold Sun" projects. Greentech Solar, November 16. http://www.greentechmedia.com/articles/read/here-comes-chinas-3b-golden-sun-projects/. Accessed 22 June 2013

Wassener B (2013) Manufacturing growth in China appears to be stabilizing. The New York Times, June 3. http://www.nytimes.com/2013/06/04/business/global/chinese-factories-appear-to-be-stabilizing.html. Accessed 19 June 2013

Wasserrab J (2013) Europe not likely to get North African electricity. Deutsche Welle, November 5. http://www.dw.de/europe-not-likely-to-get-north-african-electricity/a-16807096. Accessed 15 July 2013

Whaley F, Baardsen E, Kingsada S (2009) Hydropower helps lift rural communities. Asian Development Bank. http://www.adb.org/features/hydropower-helps-lift-rural-communities. Accessed 19 June 2013

World Resources Institute (2013) Climate analysis indicators tool 2.0. http://cait2.wri.org/wri/Country%20GHG%20Emissions?indicator=Total%20GHG%20Emissions%20Excluding%20LUCF&indicator=Total%20GHG%20Emissions%20Including%20LUCF&year=2005&act=South%20Africa&sortIdx=&sortDir=&chartType=#. Accessed 1 Aug 2013

Yahya Y (2011) World largest biofuel plant opens in Singapore. The Jakarta Globe, March 9. http://www.thejakartaglobe.com/business/worlds-largest-biodiesel-plant-opens-in-singapore/427641. Accessed 19 June 2013

Ying Y (2011) Burned by the sun. China dialogue, April 14. http://www.chinadialogue.net/article/show/single/en/4232-Burned-by-the-sun. Accessed 22 June 2013

Zadek S, Ritchken E, Fakir S, Forstater M (2010) Developing South Africa's economic policies for a low carbon world. www.zadek.net. Accessed 30 June 2013

Chapter 9
A Hybrid Model: The American Recovery and Reinvestment Act

Abstract The U.S. American Recovery and Reinvestment Act (ARRA) of 2009 offers an example of a hybrid model of EBED in which the federal government pursued both a "top-down" strategy and a "bottom-up" strategy by allocating funds to local and state entities to implement EBED programs. This chapter conducts a more in-depth review of this case, describing energy-related ARRA investments and the degree to which these investments targeted the EBED nexus. ARRA's programs, funding distributions, and the manner in which EBED initiatives were supported are explained in detail. Although the Recovery Act is too new to generate studies that evaluate the effectiveness of the various programs supported under the Act, this chapter does review the early outcomes and challenges, as accounted for in the literature to date, and extends them to the discussion of EBED initiatives at large.

The ARRA, or the "Recovery Act," was a U.S. economic stimulus package, passed during the 2007–2009 world economic recession to jump-start thousands of national and subnational programs. These programs were designed to address a range of economic, energy, social, and environmental needs, but energy and economic development investments were a significant focus within ARRA. The United States was not alone in this endeavor. China, Japan, Germany, Saudi Arabia, Italy, and other countries also adopted large-scale stimulus measures to slow the economic downturn and spark new investments in areas of national strategic importance. Across the world, these stimulus packages were often referred to as "The Green New Deal," or a means by which nations made sizable investments in the area of energy and environment to address the job loss and business contraction attributable to the global recession.

Even though the majority of funding came from the national government, ARRA is a hybrid model of EBED because it involved a range of both top-down and bottom-up activities. ARRA is an obvious example of a national-level, top-down EBED effort, because it involved a massive amount of national spending across a range of economic sectors. However, many of the programs funded under ARRA were developed or executed at the local or state level; thus, it included multiple levels of governance. Specific businesses were also part of ARRA because the legislation helped businesses develop specialized programs. We therefore highlight the inclusion of public–private partnerships in this hybrid model.

In this chapter, we provide an overview of energy-related ARRA investments and note the degree to which they focused on the EBED nexus. Next, we describe ARRA in detail with a discussion of its programs and funding distributions and the manner in which EBED initiatives were supported. Finally, we discuss the evaluation of ARRA to date. At the time of this book's writing, the Recovery Act was too new to generate studies that evaluate the effectiveness of the various programs supported under the Act. We do, however, provide a review of the early outcomes and challenges, as accounted for in the literature to date, and extend them to the discussion of EBED initiatives at large.

9.1 Overview of ARRA

In the midst of the global economic downturn, ARRA was passed on February 13, 2009, and signed into law on February 17, 2009. In the United States, the 2007–2009 recession recorded the worst post-World War II loss of employment, with 8.75 million jobs lost between January 2008 and February 2010 (Goodman and Mance 2011). Worldwide output in several sectors, as well as trade and stock market indicators, fell farther during this recession than during the Great Depression (Almunia et al. 2010).

The U.S. stimulus package, totaling approximately $840 billion, was the greatest effort since Franklin D. Roosevelt's New Deal to simultaneously target economic development and growth and technological and infrastructure development. The Recovery Act funded a series of programs focused on energy, the environment, housing, transportation, health, and public safety, among others. Given the focus of this book, it is notable that energy was a primary vehicle of programmatic operation to stimulate the economy.

The guiding objectives of the Recovery Act were to create jobs; provide relief from the recession; encourage science and health technological advances; stabilize subnational government budgets; and invest in transportation, infrastructure, and the environment (U.S. Government Printing Office 2009). The Recovery Act supported over 270,000 programs with near- or long-term goals to promote some type of economic development such as targeting growth of industries, companies, or communities or creating jobs and developing new business to help individuals or emerging firms.

In addition to the economic development focus, the U.S. energy sector was another primary target of the Recovery Act. Approximately $60 billion—plus or minus several billion, depending on which types of projects one classifies as included in this category of funding—of the total $840 billion devoted to the recovery effort was earmarked for energy-related projects. These projects focused on renewable energy, energy efficiency, smart grids, "green" jobs, and advanced fossil fuel energy.

Within these investments, the sheer diversity of EBED projects was noteworthy. Projects ranged from low-income weatherization assistance to basic R&D on algae, batteries, and solar technologies and from loan guarantees for electric vehicle battery manufacturers to smart-grid pilot projects. Much of this funding was designed to support existing energy programs, such as the Energy Efficiency Conservation Block Grant program, which is described in detail below, and the rest went toward new energy programs focused on energy planning and economic development. Above all, it is notable that the majority of energy projects under the Recovery Act fit squarely in the realm of EBED, because they included both energy *and* economic development goals.

Several other countries also adopted stimulus packages akin to ARRA. Many of these stimulus packages prominently featured policies and programs at the EBED juncture. Table 9.1 lists each country with a stimulus package between 2007 and 2009 with the total dollar value of the package, the year adopted, the amount given to energy projects, and the percentage of the total stimulus that was devoted to energy projects.

During a 2-year period, 2009–2011, over $3 trillion was invested in global stimulus funds, with approximately $463.3 billion targeted to energy and the environment. On the whole, this was only 15 % of the total stimulus funds, but the variance between countries was wide, masking some of the significant investments made by some countries in EBED. Countries that devoted more than 20 % of their stimulus to energy and the environment include Australia (21 %), France (21 %), China (33 %), and South Korea (95 %). The European Union committed approximately 59 % of its funds to energy and the environment. It is important to note, however, that these figures may include activities that are slightly outside of the EBED realm, because not all environmental activities or energy activities fit within the EBED domain. Approximately two-thirds of the U.S. estimate for funding devoted to energy and the environment, for example, is what we would identify as EBED funding, while the other third is not.

9.2 Energy-Related Recovery Act Offices and Programs

Out of approximately 270,000 ARRA projects from the contracts, grants, and loans awarded in the United States, approximately 8,600 programs were energy related. Figure 9.1 displays the awarding agencies responsible for these energy-related programs. The Departments of Energy (DOE) and Transportation managed the lion's share of these ARRA programs. We describe their funding, program allocations, and operations to detail how a nationally driven initiative of this magnitude administers these resources into relevant programs.

Energy-related ARRA programs and projects, as categorized here, include all ARRA-funded activities related to renewable energy, energy conservation, electrical transmission and distribution, nondefense energy program clean-up, energy-related housing retrofits, high-speed rail and intercity passenger rail, and other

Table 9.1 Stimulus funding in different counties

	Total stimulus (US$ billion)	Year stimulus bill adopted[a]	Energy- and environmental-related stimulus (US$ billion)	Energy and environment spending as % of total stimulus
Argentina	13.2	2008		
Australia	43.8	2008	9.3	21.2
Brazil	3.6	2008		
Canada	31.8	2009	2.8	8.3
China	647.5	2008	216.4	33.4
France	33.7	2008	7.1	21.2
Germany	104.8	2009	13.8	13.2
India	13.7	2008		
Indonesia	5.9	2009	0.1	1.7
Italy	103.5	2009	1.3	1.3
Japan	639.9	2009	36.0	5.6
Mexico	7.7	2009	0.8	9.7
Russia	20.0	2008		
Saudi Arabia	126.8	2008	9.5	7.5
South Africa	7.5	2008	0.8	10.7
South Korea	38.1	2008	36.3	95.2
United Kingdom	34.9	2008	3.7	10.6
United States[b]	787.0	2009	94.1	12.0
European Union[c]	38.8	2008	22.8	58.7
Global total	3,016.3		463.3	15.4

Source This table and footnotes adopted from Barbier (2010) and Robins et al. (2009)
[a] Most countries passed their stimulus packages in waves that came in both late 2008 and early 2009. We used the following criteria when choosing an adoption year: when one wave was more closely or prominently (i.e., in the press) associated with energy or environmental stimulus funding, we selected the year for that wave; when energy or environmental stimulus was not a significant factor, but one wave was greater in dollar terms, we selected the larger
[b] From the February 2009 ARRA only. The October 2008 Emergency Economic Stabilization also included U.S.$185 billion in tax cuts and credits, including US$18.2 billion for investments in wind, solar, and carbon capture and storage. The value provided in this table was the original estimate for ARRA and has since been revised to $840 billion, in keeping with the President's 2012 budget
[c] Only the direct contribution by the European Union is included

energy-related transportation funding. Figure 9.2 displays the energy-related programs that received the greatest funding. The biggest EBED recipient programs included DOE's Loan Guarantee Program, grants for alternative transportation, Federal Transit Formula Grants, and the Weatherization Assistance Program.

Recovery Act funds were dispersed through five different mechanisms: formula, competition, demand from certain entities, one-time allocations, or agency spending plans (Fox et al. 2009). Table 9.2 demonstrates the variation in allocation method across programs for DOE. All awards allocated through DOE are presented by recipient program, amount, and allocation method. As this table

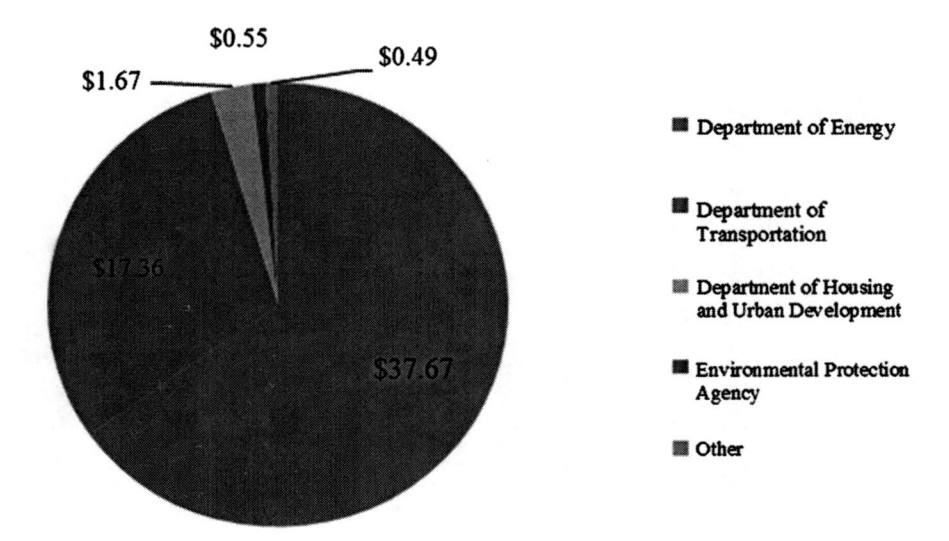

9.2 Energy-Related Recovery Act Offices and Programs

Fig. 9.1 Energy-related contract, grant, and loan funds awarded under ARRA 2009 as of fourth quarter calendar year 2011 for the four highest-awarding agencies. Values displayed are in billions of dollars. *Source* Recovery Accountability and Transparency Board 2012

demonstrates, ARRA included a variety of different types of policy instruments, many of which are reviewed in Chap. 4, and these policy instruments were applied in a range of different programs.

The Recovery Act did not just funnel resources through the U.S. DOE; it involved every federal agency in some way. In Sect. 9.2.1, we provide an overview of four different offices within federal government agencies that managed EBED programs and give examples of some of the programs under the purview of these offices that demonstrate EBED characteristics.

9.2.1 Office of Energy Efficiency and Renewable Energy

The Office of Energy Efficiency and Renewable Energy (EERE) is a program within DOE that is responsible for developing and deploying energy efficient and renewable energy technologies. EERE was provided approximately $16.8 billion of ARRA funding to implement 20 different programs that targeted advanced transportation, energy efficiency, and renewable energy electricity generation investments. EERE programs funded under ARRA included the following: Biomass, Building Technologies, Community Renewable Energy Deployment, Federal Energy Management, Weatherization Assistance Program, the Energy Efficiency and Conservation Block Grants, and the Wind Energy Technology Program. Two of these programs, the Weatherization Assistance Program and the

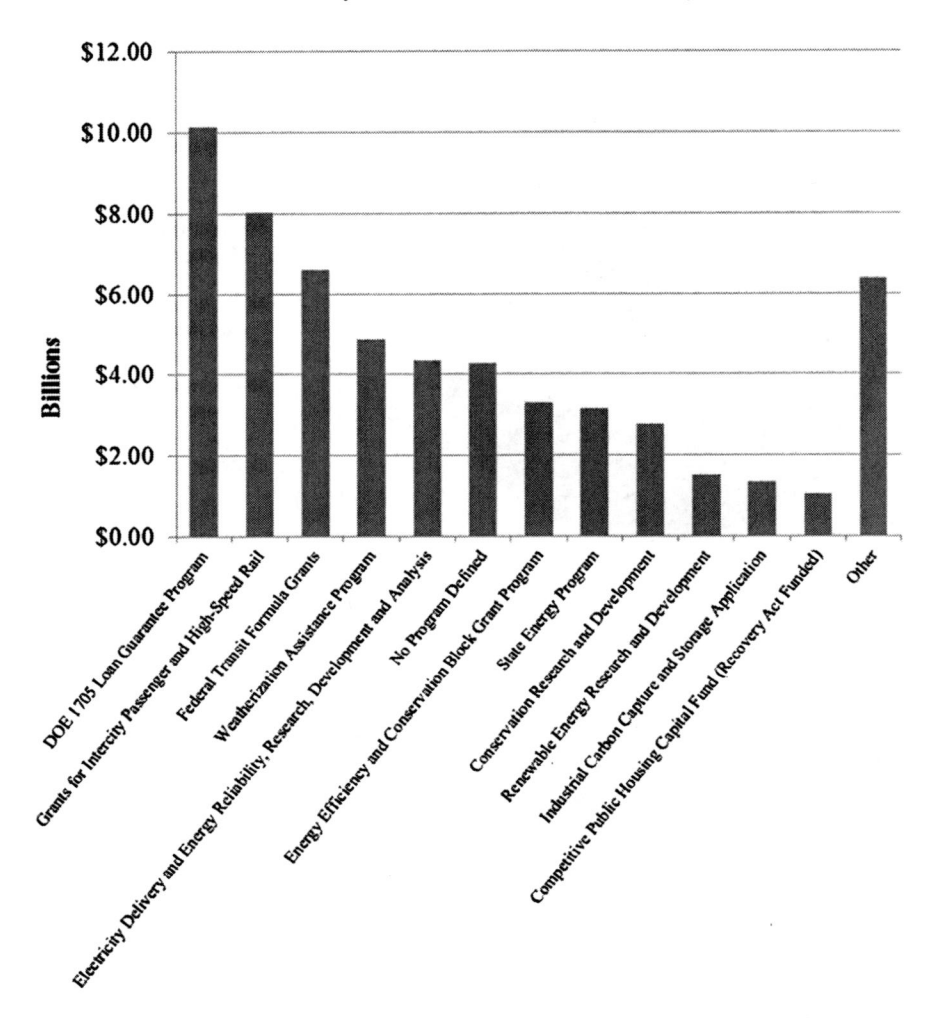

Fig. 9.2 Energy-related contract, grant, and loan funds awarded under ARRA 2009 by Program as of fourth quarter calendar year 2011. *Source* Recovery Accountability and Transparency Board 2012

Energy Efficiency and Conservation Block Grants programs, prominently featured EBED activities through ARRA. For these reasons we describe their efforts in more detail.

The Weatherization Assistance Program, which existed prior to the passage of the Recovery Act, received approximately $4.98 billion in ARRA funding. Through financing and support to 900 local agencies, the program provided energy efficiency investment—up to $6,500 per eligible household—and, consequently, energy savings for low-income families. The Weatherization Assistance Program

9.2 Energy-Related Recovery Act Offices and Programs

Table 9.2 DOE program allocations

Recipient program	Amount (U.S.$ million)	Allocation method
Energy Efficiency and Renewable Energy	3,500	Agency discretion: staffing and operational expenses (U.S. ARRA) (U.S. Government Printing Office 2009)
Energy Efficiency and Conservation Block Grants	3,200	$2800 million by population; $400 million by competitive award (42 U.S.C 17153) (U.S. House of Representatives, The Office of the Law Revision Counsel 2011c)
Weatherization Assistance Program	5,000	State applications, noncompetitive (42 U.S.C. 6864) (U.S. House of Representatives, The Office of the Law Revision Counsel 2011d)
State Energy Program	3,100	State applications, noncompetitive (10 C.F.R. 420 per 42 U.S.C. 6321) (U.S. House of Representatives, The Office of the Law Revision Counsel 2011a)
Manufacturing of Advanced Batteries and Components	2,000	Competitive awards (42 U.S.C. 17011 et seq.) (U.S. House of Representatives, The Office of the Law Revision Counsel 2011e)
Electricity Delivery and Energy Reliability	4,310	Competitive awards (U.S. DOE DE-FOA-0000058 and DE-FOA-0000098) (U.S. DOE, Office of Electricity Delivery and Energy Reliability 2009; U.S. DOE, National Energy Technology Laboratory 2009a)
Worker training activities	100	Competitive awards (U.S. DOE DE-FOA-0000152) (U.S. DOE, Richland Operations Office 2009)
Resource assessment and an analysis of future demand and transmission requirements	60	Competitive awards (U.S. DOE DE-FOA-0000068 and Office of Electricity Delivery and Energy Reliability) (U.S. DOE, National Energy Technology Laboratory 2009b; U.S. DOE, Office of Electricity Delivery and Energy Reliability 2009)
Additional funding for smart-grid research	10	Direct funding of existing functions (U.S. DOE, Office of Electricity Delivery and Energy Reliability 2010)
Fossil energy research and development	3,400	Competitive awards; private and public recipients, including universities (U.S. DOE, Office of Fossil Energy 2009)

(continued)

Table 9.2 (continued)

Recipient program	Amount (U.S.$ million)	Allocation method
Nondefense environmental cleanup	483	Direct funding of existing functions (U.S. DOE, Office of Environmental Management 2009b)
Uranium enrichment decontamination and decommissioning	390	Direct funding of existing functions (U.S. DOE, Office of Environmental Management 2009c)
Science	1,600	Direct funding of existing functions (U.S. DOE, Office of Science 2009)
Advanced Research Projects Agency—Energy (ARPA-E)	400	Competitive awards (U.S. DOE, Advanced Research Projects Agency—Energy 2010)
Innovative Technology Loan Guarantee Program	6,000	Competitive awards (U.S. DOE, Loan Programs Office 2009)
Office of the Inspector General	15	Direct funding of existing functions (U.S. ARRA) (U.S. Government Printing Office 2009)
Defense environmental cleanup	5,127	Direct funding of existing functions (U.S. DOE, Office of Environmental Management 2009a)
Western area power administration	10	Direct funding of existing functions (42 U.S.C. 7152) (U.S. House of Representatives, The Office of the Law Revision Counsel 2011b)

funds were allocated based on the number of low-income households in a state, local heating and cooling degree-days, and low-income residential energy costs. Funds were dispersed incrementally to ensure weatherization progress and quality control, where recipients received the first 10 % to begin operations, the next 40 % after submitting plans for weatherizing homes, and the remaining funds after weatherizing 30 % of their homes and performing certain oversight tasks (U.S. Government Accountability Office 2011). As of February 2011, 377,655 homes were weatherized nationwide from ARRA support (U.S. DOE, Weatherization Assistance Program 2011).

The Energy Efficiency and Conservation Block Grants program was created by the Energy Independence of Security Act of 2007 but first funded by ARRA, which devoted $3.2 billion in formula and competitive grants to subnational governments for a variety of energy efficiency, job creation, and manufacturing projects (U.S. DOE 2010). Activities under this program include efficiency and conservation strategies, technical consultant services, energy audits, building and other energy efficiency retrofits, financial incentive programs, and building and facility conservation programs, among others.

9.2.2 Office of Electricity Delivery and Energy Reliability

The Office of Electricity Delivery and Energy Reliability within DOE is responsible for improving the nation's electric grid to make it more efficient and reliable. The Office's objectives are to upgrade the nation's power grid, create a more secure and reliable transmission system, increase grid efficiency, allow buyers and sellers to access electricity data, increase demand responsiveness, reduce emissions associated with the electricity transmission and distribution, and facilitate the penetration of alternative energy technologies onto the transmission grid.

The Office was allocated $4.5 billion in ARRA funding to help advance electric grid and technology research and management (U.S. DOE 2010). Approximately $3.5 billion of ARRA funding for this office was used to match grants with utilities and other companies to improve the smart-grid infrastructure provided by utilities and other entities. This office also used ARRA to leverage an existing program, Section 1304s Smart Grid Regional and Energy Storage Demonstration Projects program, to build 32 demonstration projects. Half of these projects focused on developing smart grids, and half were devoted to advancing our understanding about ways to improve energy storage.

Workforce development was also supported through the Office of Electricity Delivery and Energy Reliability's ARRA programs. The program trained the labor force to install and work with smart grids, as well as trained specific subsectors such as veterans, the unemployed, and utility workers from the old grid economy (U.S. DOE 2010).

9.2.3 DOE Loan Programs Office

The DOE's Loan Guarantee Program was originally established under the Energy Policy Act of 2005 with the mission to provide loans to commercially advance new and promising technologies. The Recovery Act added $3.94 billion in loan support to help launch emerging technology and commercial-scale renewable and transmission technologies under the Innovative Technology Loan Guarantee Program (also known as the Section 1705 Loan Program) (U.S. DOE 2010; U.S. DOE, Loan Programs Office 2013). Of this amount $25 million was reserved for administrative costs and $10 million for the Advanced Technology Vehicle Manufacturing program, which provided financial backing to qualified manufacturers of alternative vehicles. The 30 companies that received support from the Section 1705 Loan Program fit into one of five categories (with total number of companies that received loan support in parenthesis): cell, battery, and materials manufacturing facilities (9); advanced battery supplier manufacturing facilities (10); advanced lithium-ion battery recycling facilities (1); electric drive component manufacturing facilities (7); and electric drive subcomponent manufacturing facilities (3) (Allison 2012). The remaining funds were devoted to a variety of renewable energy, electric transmission systems, and biofuel loan opportunities. The 24 loans made in this program targeted biofuels), energy storage, geothermal, solar manufacturing, solar generation, transmission, and wind generation (Allison 2012). The first conditional commitment from the Section 1705 Loan Program fund was made to Solyndra, Inc.[1] in March 2009 and, by February 2012, there were 26 closed loans under the 1705 program and five closed loans under the vehicle manufacturing program, for a total of $35.1 billion in guaranteed loans (U.S. DOE, Loan Programs Office 2012).

9.2.4 Department of Housing and Urban Development's Green Retrofit Program

The Green Retrofit Program, under the Department of Housing and Urban Development (HUD), was allocated $250 million of ARRA funds to provide loans, grants, and other incentives for green building retrofits of multifamily housing units, including homes owned by low-income, elderly, and disabled residents (U.S. HUD 2011). Approximately $155 million went toward grants and $80 million toward loans, with the remaining funds devoted to training, monitoring, and evaluation. Upon acceptance of their applications, grant and loan recipients were required to undergo third-party assessments of their properties to determine retrofit potential; projects could receive a maximum of $15,000 per unit. This retrofit program was designed to help conserve energy and water; reduce GHG emissions; improve indoor air quality and comfort; reduce costs to the homeowner; create jobs; and train residents in property maintenance, green building, and conservation (U.S. HUD 2010). No post-2010 statistics are readily available from HUD as to the number of homes served, although program metrics include funds obligated by HUD, funds expended by grantees, how many units were rehabilitated, and how many units were retrofitted (U.S. HUD 2010).

---

[1] Solyndra, Inc. applied for a loan under the DOE Section 1705 Loan Program in December 2006 and, along with 15 other firms, was invited to submit a complete loan application in October 2007. Solyndra submitted its application in August 2008, and a credit review board reviewed the application in January and again in March 2009. The application was conditionally approved, given that Solyndra could raise an additional $198 million on top of the DOE's $535 million loan. The loan was approved and Secretary of Energy Chu announced the loan in person at Solyndra's headquarters in September 2009. By March 2010, Solyndra demonstrated signs of financial hardship. The DOE notified Solyndra that it needed to raise additional capital before the next loan disbursement. The DOE loaned Solyndra an additional $75 million in February 2011 but did not commit to a subsequent loan of an additional $75 million. Solyndra announced bankruptcy at the end of August 2011. For additional details on the Solyndra case, refer to U.S. House Committee on Energy and Commerce memos (U.S. House Committee on Energy and Commerce, Subcommittee on Oversight and Investigations Staff 2011a, b; U.S. House Committee on Energy and Commerce, Subcommittee on Oversight and Investigations Democratic Staff 2011c).

9.2.5 Commonalities Within Programs

A couple of common elements unite all four of these offices and the variety of programs offered by each. First, each of these programs squarely fits into the EBED framework, with an integration of energy planning and economic development goals. Second, each program was aligned and supported by a well-established federal agency. The ARRA-funded program mission was consistent with the agencies' standard operations. Third, each program involved a variety of stakeholders, including federal bureaucrats but also subnational actors as well, including state or local governments, homeowners, representatives from corporate industries, academics and other researchers, and local nonprofits or other organizations. Fourth, each program was supported by a mix of different policy instruments. Many of these programs targeted energy technology innovation and adoption, workforce training, and industrial growth strategies. Finally, each program adopted its own evaluation metrics, although it is still too early to experience the full effects and impacts of these programs—with the exception of the Weatherization Assistance Program, which has published numbers of total houses weatherized to date. DOE measured outcomes from the types of programs reviewed above based on obligations and expenditures, tallies of installed hardware or capacity, and measures of energy savings. The DOE Loan Guarantee Program evaluation metrics included measures of financing or operation, GHG savings, and electricity generation or manufacturing capacity. The Office of Electricity Delivery and Energy Reliability used a set of metrics to evaluate its smart-grid deployments as well as their economic and environmental impacts, including measures of the amount of hardware deployed, energy and cost savings, and pollution reduction. HUD's evaluation metrics for their Green Retrofit Program include cost and energy savings, health effects, and fiscal sustainability.

9.3 Case Studies of Selected Local Recovery Act–Funded Initiatives

Many case studies of local initiatives sponsored by ARRA funds fit in the EBED realm, two of which are outlined below. Because these cases are subnational in nature, they are similar to the cases we discuss in Chap. 7.

9.3.1 Green Launching Pad

The Green Launching Pad is an ongoing energy technology commercialization program that was funded through ARRA. This business incubator seeks to provide funding for entrepreneurs in the state of New Hampshire to help them accelerate the commercialization process for renewable energy, energy efficiency, conservation, and emissions reduction technologies. Green Launching Pad hits squarely on the energy and economy juncture in its mission to spur innovation and entrepreneurship in low-emissions, advanced, and efficient energy. This initiative represents a market-based approach to EBED, in which the incubator supports the development and growth of new businesses.

The Green Launching Pad offers a competitive grant program for entrepreneurs who propose new business plans with energy or climate objectives. With accelerated business growth, it also brings economic development benefits such as new jobs, increased revenues, and technology-based industry growth within the state of New Hampshire. The initiative is run by a partnership of representatives from the New Hampshire Governor's Office, the State Office of Energy and Planning, the University of New Hampshire, and the New Hampshire Charitable Foundation.

The initiative offers grants to successful teams of applicants that range from $20,000 to $90,000 in business development through technical and professional services. Once selected for the program, successful applicants enroll in an intensive summer program that is guided by the Green Launching Pad Working Group. At the end of the intensive program, the entrepreneurial teams present their final business plan to public and private investors.

Technologies or other business pursuits that were funded as of early 2012 include biomass power generation, CO_2 reduction technologies, organic semiconductors, industrial energy efficiency, financing solutions, energy storage, revitalization of hydropower in former industrial mill towns, self-contained solar-powered LED traffic signals, sustainable footwear manufacturing, solar collectors, and a hydrokinetic turbine generation system. This list demonstrates a range of technologies on which business models have been built and commercialization paths better established as a result of this program.

9.3.2 Energize Phoenix

The Energize Phoenix initiative provided financial assistance with energy efficiency measures in a corridor of neighborhoods that span a 10-mile stretch of the Phoenix light rail line. Energize Phoenix, which is also an ongoing initiative, was funded by a $25 million grant from the U.S. DOE Better Buildings Program and ARRA. These funds helped the program target residential homeowners, business owners, and other property owners with rebates, grants, and revolving loan funds to help them

- reduce home energy consumption by up to 30 %,
- reduce commercial energy use by up to 18 %,
- reduce carbon emissions by as much as 50,000 metric tons per year, and
- provide energy efficiency upgrades for approximately 1,700 homes and 30 million square feet of office and industrial space.

9.3 Case Studies of Selected Local Recovery Act–Funded Initiatives

The initiative also aims to educate property owners about energy efficiency and energy use and to provide jobs in the Phoenix area. This initiative, therefore, represents an approach that is a mix of household and place based. Targeted beneficiaries of the initiative are households and individuals; households will benefit because their income will increase and their household energy consumption will decrease and specific neighborhoods will experience revitalization and economic growth.

The residential component of this initiative includes three programs: a rebate match program, a homeowner grant program, and a rental housing grant program. The rebate match program is available to all residential homeowners and guaranteed that Energize Phoenix would match any rebates offered through local programs for energy efficiency upgrades, including heating and cooling, duct leakage repair, sealing, insulation, shade screens, or solar water heater upgrades. The homeowner grant program is available for applicable low- to moderate-income households and covers up to 60 % of the total energy efficiency upgrade costs.

The rental housing grant program is somewhat similar but is available to rental property owners that serve low- to moderate-income tenants with specific income thresholds. To qualify, tenants must also comply with accessibility, historic preservation, and housing quality standards. The grants may not exceed $3,000 per unit or the cost to achieve an estimated 15 % energy savings. Further, if the property continues under the same ownership, 10 % of the conditional grant is forgiven annually.

The business program targets schools and businesses with similar energy efficiency improvements in the Energize Phoenix corridor. The initiative focuses on lighting, refrigeration, and HVAC upgrades and works in coordination, both technically and financially, with Arizona Public Service (APS) Express Solutions, a utility company. The costs of efficiency investments that are not covered by rebates from APS Express Solutions and the Energy Phoenix initiative are eligible for loans supplied by a revolving loan fund program.

In addition to these three components, Energize Phoenix includes a subprogram called the Energy Dashboard Program, which helps property owners monitor and evaluate their energy use, both in aggregate and by each appliance. The coordinators of this program, Arizona State University and the City of Phoenix's Neighborhood Services Department, are also able to monitor end users' energy use and conduct analyses on the effectiveness of energy efficiency information campaigns targeted at these residents.

The Energize Phoenix initiative is administered by a partnership between Arizona State University, APS, and the City of Phoenix, including the Public Works, Neighborhood Services, and Community Economic Development units. The National Bank of Arizona also participates in the revolving loan fund program.

9.3.3 Summary of Case Studies

These two cases reveal several common themes. First, each initiative involves a diverse group of program facilitators, including representatives from the public and private realms. In many of these cases, this diversity of actors is necessary given the cross-disciplinary nature of these initiatives, which target a number of different actors, such as homeowners, business owners, entrepreneurs, and employees. Second, in the Energize Phoenix case, public ARRA funds were used to leverage additional financial support from other parties. The case of the accelerated business incubator, the Green Launching Pad, also highlights a similar trend: public funds are used to attract additional fiscal support, in this case from investors of start-up businesses. Third, these programs have developed evaluation programs or metrics to ensure that the program is as effective and efficient as possible. Energize Phoenix, for example, has an evaluation team comprising collaborating institutions.

9.4 Early Evaluations of ARRA and Potential Implications

Although the Recovery Act was passed within a few years of this writing, several analysts have documented early effects and challenges associated with the programs. In keeping with the federal government's adherence to transparency and accountability, the U.S. government has produced a substantial number of assessments of Recovery Act programs, which outline program impacts, difficulties with implementation, and suggestions for refinement and improvement. Recipients of U.S. stimulus funds reported their progress quarterly, and these reports are available for public viewing at the website recovery.org. The main themes that have emerged from these reports, including those conducted by independent analysts as well, include positive reviews of economic and energy-related program outcomes but also some concerns with federal, state, and local administrative capacity, monitoring and reporting issues, and the pace at which these programs were successfully implemented.

The early impacts of ARRA have been tracked by GDP, job growth, and energy savings measures. One study estimated the effects of ARRA based on funding notifications to be between 4.3 and 8.3 jobs created or saved per state for each $1 million in spending (Wilson 2011). The Congressional Budget Office (2011) estimated that real GDP rose 0.3–1.9 % in the third quarter of 2011 due to the Recovery Act (U.S. Congressional Budget Office 2011). Another study found that $93 billion in green economic investments from ARRA, including but not limited to energy programs, increased GDP by $146 billion and added or maintained 997,000 direct, indirect, and induced jobs (Walsh et al. 2011). Two early studies have estimated energy savings from ARRA programs. Pierce (2011) found an ARRA-induced reduction in the Department of Defense's energy use, which saved money and increased energy security in ways that both directly and indirectly saved the lives of troops on the battlefield. Eisenberg (2010) found that ARRA support of the weatherization assistance program saved each household approximately 29 million BTU of energy, 2.65 metric tons of carbon dioxide, and between $104 and $174 in avoided energy costs.

Other studies suggest that energy-related ARRA funds significantly helped the renewable energy and energy efficiency industries by providing financing during a time of financial stress and when energy project developers otherwise might have had a difficult time securing funding (Hargreaves 2010). Financial support for energy investments from the Recovery Act also leveraged additional private and public funding. According to state officials who participated in a 2010 Environmental and Energy Study Institute panel, for every $1 of ARRA funding for energy projects, states and private sources contributed $10.71 (EESI 2010).

These successes, however, were despite several challenges. Early reports indicate that, in some cases, ARRA funding outpaced the ability of the agencies to implement even their most "shovel-ready" projects, a term used to describe projects that were immediately ready upon injection of funds. Actual ARRA funding levels were much lower than planned (Johnson 2009). For example, as of early 2012, by which time most ARRA funding was intended to be spent, only 60 % of total DOE funding was actually spent even although 98 % of it was allocated (U.S. DOE, Office of Inspector General 2012). The reason for this lag between funding allocation and actual spending is primarily attributable to three factors. First, a general lack of staff capacity at multiple levels of government impeded the progress of ARRA projects. In addition to the significant pressures of administering large amounts of funding quickly, inadequate capacity to administer programs was also due to furloughs from the 2007 economic downturn. Both of these factors made it difficult to implement ARRA projects in a timely manner and to ramp up efforts at a pace consistent with the injection of ARRA funds (Johnson 2009; U.S. Government Accountability Office 2011). This finding is further evidenced by the several locations needing to add staff and administrative structures to cope with new ARRA procedures (Wyatt 2009). The DOE in particular was noted in several government reports as struggling to maintain a workforce that could keep up with ARRA needs (U.S. DOE, Office of Inspector General 2012).

Second, the concept of "shovel-ready" or even 2-year implementation in the context of energy projects may be an oxymoron, where energy projects often require long-term, consistent funding streams and these projects may take several years to implement. Third, the administrative burdens associated with the government's transparency and reporting requirements, compliance with preexisting laws such as the National Environmental Policy Act, and compliance with additional ARRA-specific requirements such as "buy American" mandates added significantly to the burden of spending ARRA funds in a timely and efficient way (U.S. Government Accountability Office 2011).

Over the coming years, we expect a number of publications that document energy-related ARRA outcomes and impacts. It is important to note, however, that not all ARRA impacts will be realized and measureable in the short term, as is commonly the case with science, technological innovation, and infrastructural investments.

9.5 Conclusions

ARRA provides a complex example of an EBED approach that is funded by the national government but requires extensive local implementation, with necessary coordination between economic development and energy program officials. ARRA programs focused heavily on energy technologies and the use of these technologies to generate economic development and growth opportunities. The energy technologies and services included in the ARRA programs reviewed in this chapter represent a wide range but are all advanced, efficient, and low emissions.

ARRA programs demonstrate a mix between longer-term investments in technology innovation and industry development and shorter-term investments in economic stimulus within energy and related sectors. This blend of innovation, low-income assistance, and industry-targeted stimulus is truly a multifaceted approach to energy and economic development.

The programs created or expanded through ARRA also represented a range of EBED approaches. Some programs were distinctly market based, such as those that supported electric vehicle and battery technologies or smart-grid infrastructure development. Other programs assumed a household-based approach, such as weatherization programs, or a place-based approach, such as programs that targeted specific cities or neighborhoods. Some of the more comprehensive initiatives that ARRA supported, such as the Energize Phoenix and Green Launching Pad programs reviewed above, use a combination of market-, household-, and place-based approaches. All of the programs and offices reviewed in this chapter involved multiple actors from various levels of government, the private sector, and a variety of other organizations and individuals. Many of these individuals or entities may not have focused on EBED activities before ARRA, but rather extended their purview to include EBED projects once the legislation was passed and funding was in place.

Despite the reports of positive ARRA outcomes, the programs under ARRA also encountered several challenges. Some of these challenges were likely inevitable given the sheer scale of the grand ARRA experiment and the time pressures to funnel billions of dollars into local economies quickly with a high degree of transparency and documentation. Other challenges surfaced within state and local governments that were at times caught ill-prepared because of stresses from the recession and inexperience in handling large-scale projects in short time horizons. These challenges highlight the importance of timing, a strong institutional capacity, and realistic expectations in EBED initiatives. Although all such challenges with ARRA implementation and constructive lessons have yet to be identified, these preliminary findings about ARRA can lend insights to those that operate within the EBED domain.

References

Allison H (2012) Report of the independent consultant's review with respect to the Department of Energy Loan and Loan Guarantee Portfolio. The White House, Washington

Alumnia M, Bénétrix A, Eichengreen B, O'Rourke KH, Rua G (2010) From Great Depression to great credit crisis: similarities, differences and lessons. Econ Policy 25:219–265

Barbier EB (2010) A global green new deal: rethinking the economic recovery. Cambridge University Press, New York

Eisenberg JF (2010) Weatherization assistance program technical memorandum: background data and statistics. ORNL/TM-2010/66. Oak Ridge National Laboratory, Oak Ridge

Environmental and Energy Study Institute (EESI) (2010) Economic impacts of Recovery Act funding for the state energy program. Environmental and Energy Study Institute, Washington. http://www.eesi.org/economic-impacts-recovery-act-funding-state-energy-program-07-jul-2010. Accessed 8 March 2012

Fox R, Walsh J, Fremstad S (2009) Bringing home the green recovery: a user's guide to the 2009 American Recovery and Reinvestment Act. PolicyLink and Green for All, Oakland

Goodman CJ, Mance SM (2011) Employment loss and the 2007–09 recession: an overview. Mon Labor Rev 134:3–12

Hargreaves S (2010) The stimulus project: how stimulus saved renewable energy. Cable News Network, Jan 26. http://money.cnn.com/2010/01/24/news/economy/stimulus_wind/index.htm?postversion=2010012416

Johnson N (2009) Does the American Recovery and Reinvestment Act meet local needs? State Local Gov Rev 41:123–127

Pierce BJ (2011) A new shade of camouflage: the American Recovery and Reinvestment Act helps the Department of Defense go green. Social Science Research Network. http://ssrn.com/abstract=1967138 or http://dx.doi.org/10.2139/ssrn.1967138. Accessed 12 April 2012

Recovery Accountability and Transparency Board (2012) Recovery explorer. http://www.recovery.gov/Transparency/Pages/DataExplorerLanding.aspx. Accessed 21 Feb 2011

Robins N, Padamadan R, Clover R (2009) More green money on the table. HSBC Global Research, London

U.S. Congressional Budget Office (2011) Estimated impact of the American Recovery and Reinvestment Act on employment and economic output from July 2011 through September 2011. U.S. Congressional Budget Office, Washington

U.S. Department of Energy (DOE) (2010) Revised DOE Recovery Act plan—June 2010. U.S. Department of Energy, Washington

U.S. Department of Energy (DOE), Advanced Research Projects Agency—Energy (ARPA-E) (2010) ARPA-E program specific recovery plan. U.S. Department of Energy, Washington

U.S. Department of Energy (DOE), Loan Programs Office (LPO) (2009) Innovative technology loan guarantee program specific recovery plan. U.S. Department of Energy, Washington

U.S. Department of Energy (DOE), Loan Programs Office (LPO) (2012) Our projects. http://lpo.energy.gov/our-projects/. Accessed 20 Nov 2011

U.S. Department of Energy (DOE), Loan Programs Office (LPO) (2013) About the Section 1705 loan program. http://lpo.energy.gov/programs/1705-2/. Accessed 24 July 2013

U.S. Department of Energy (DOE), Office of Electricity Delivery and Energy Reliability (OE) (2009) Smart Grid Investment Grant program. Funding opportunity announcement, DE-FOA-0000058. U.S. Department of Energy, Washington

U.S. Department of Energy (DOE), Office of Electricity Delivery and Energy Reliability (OE) (2010) Recovery program plan. U.S. Department of Energy, Washington

U.S. Department of Energy (DOE), Office of Environmental Management (EM) (2009a) Defense environmental cleanup program specific recovery plan. U.S. Department of Energy, Washington

U.S. Department of Energy (DOE), Office of Environmental Management (EM) (2009b) Nondefense environmental cleanup program specific recovery plan. U.S. Department of Energy, Washington, DC U.S. Department of Energy, Washington

U.S. Department of Energy (DOE), Office of Environmental Management (EM) (2009c) Uranium decontamination and decommissioning program specific recovery plan. U.S. Department of Energy, Washington

U.S. Department of Energy (DOE), National Energy Technology Laboratory (NETL) (2009a) Local Energy Assurance Planning (LEAP) Initiative. Funding opportunity announcement, DE-FOA-0000098. U.S. Department of Energy, Washington

U.S. Department of Energy (DOE), National Energy Technology Laboratory (NETL) (2009b) Recovery Act-resource assessment and interconnection-level transmission analysis and planning. Funding opportunity announcement, DE-FOA-0000068. U.S. Department of Energy, Washington

U.S. Department of Energy (DOE), Office of Fossil Energy (FE) (2009) Fossil energy research & development (R&D) program specific recovery plan. U.S. Department of Energy, Washington

U.S. Department of Energy (DOE), Office of Inspector General, Office of Audits and Inspections (2012) Lessons learned/best practices during the Department of Energy's implementation of the American Recovery and Reinvestment Act of 2009. OAS-RA-12-03. U.S. Department of Energy, Washington

U.S. Department of Energy (DOE), Office of Science (SC) (2009) Office of Science program specific recovery plan. U.S. Department of Energy, Washington

U.S. Department of Energy (DOE), Richland Operations Office (2009) Workforce training for the electric power sector. Funding opportunity announcement, DE-FOA-0000152. U.S. Department of Energy, Washington

U.S. Department of Energy (DOE), Weatherization Assistance Program (2011) Homes weatherized by grantee in February 2011 (Calendar Year). U.S. Department of Energy, Washington

U.S. Department of Housing and Urban Development (HUD) (2010) Revised agency wide program plan—June 17, 2010. Department of Housing and Urban Development, Washington. http://portal.hud.gov/hudportal/documents/huddoc?id=T29_DOC_105.pdf. Accessed 24 July 2013. U.S

U.S. Department of Housing and Urban Development (HUD) (2011) HUD Implementation of the Recovery Act. http://portal.hud.gov/hudportal/HUD?src=/recovery/about. Accessed 20 Nov 2011

U.S. Government Accountability Office (2011) Recovery Act: energy efficiency and conservation block grant recipients face challenges meeting legislative and program goals and requirements. GAO-11-379. U.S. Government Accountability Office, Washington

U.S. Government Printing Office (2009) American Recovery and Reinvestment Act of 2009. U.S. Government Printing Office, Washington

U.S. House Committee on Energy and Commerce, Subcommittee on Oversight and Investigations Staff (2011a) RE: Hearing on "Solyndra and the DOE Loan Guarantee Program." Memo to members, Subcommittee on Oversight and Investigations. U.S. House Committee on Energy and Commerce, Washington

U.S. House Committee on Energy and Commerce, Subcommittee on Oversight and Investigations Staff (2011b) RE: Hearing on "the Solyndra failure: views from DOE Secretary Chu." Memo to members, Subcommittee on Oversight and Investigations. U.S. House Committee on Energy and Commerce, Washington

U.S. House Committee on Energy and Commerce, Subcommittee on Oversight and Investigations Democratic Staff (2011c) Re: Hearing titled "From DOE Loan Guarantee to bankruptcy to FBI raid: what Solyndra's executives knew." Memo to Democratic members, Subcommittee on Oversight and Investigations. U.S. House Committee on Energy and Commerce, Washington

U.S. House of Representatives, The Office of the Law Revision Counsel (2011a) 10 C.F.R. 420 per 42 U.S.C. 6321

U.S. House of Representatives, The Office of the Law Revision Counsel (2011b) 42 U.S.C. 7152
U.S. House of Representatives, The Office of the Law Revision Counsel (2011c) 42 U.S.C 17153
U.S. House of Representatives, The Office of the Law Revision Counsel (2011d) 42 U.S.C. 6864
U.S. House of Representatives, The Office of the Law Revision Counsel (2011e) 42 U.S.C. 17011 et seq
Walsh J, Bivens J, Pollack E (2011) Rebuilding green: the American Recovery and Reinvestment Act and the green economy. BlueGreen Alliance and Economic Policy Institute, Washington
Wilson DJ (2011) Fiscal spending jobs multipliers: evidence from the 2009 American Recovery and Reinvestment Act. Working Paper 2010-17. Federal Reserve Bank of San Francisco, San Francisco
Wyatt DFG (2009) The perceived challenges of implementing the American Recovery and Reinvestment Act. State Local Gov Rev 41:128–132

For Additional Reading on the Case Studies

American Council for an Energy-Efficient Economy (2011) Case study: clean energy works Portland. American Council for an Energy-Efficient Economy, Washington
Boss S (2010) This old green house. Stanford Soc Innov Rev 8:61–62
City of Portland, Oregon (2009) Community workforce agreement on standards and community benefits in the Clean Energy Works Portland Pilot Project
Dalrymple M, Bryck D (2011) Energy efficiency on an urban scale—year one report: from the ground up. In: Heffernon R (ed) Arizona State University Global Institute of Sustainability, Tempe
Energize Phoenix (2010) Energize Phoenix. http://www.energizephx.com
Fraser M (2010) Energize Phoenix: transformation through behavior and retrofits along the green rail corridor. In: Paper presented at the Comparative Genetics of Cities, University College London, 2010
Green for All (2009) Recovery innovation in Portland. Communities of practice. [Podcast] http://www.google.com/url?sa=t&rct=j&q=&esrc=s&source=web&cd=1&ved=0CC0QFjAA&url=http%3A%2F%2Fwww.portlandonline.com%2Fshared%2Fcfm%2Fimage.cfm%3Fid%3D275721&ei=aQXwUZLRD9bh4AOOvIHAAw&usg=AFQjCNGz760fSDFLJ1Wq7p7P78hea09cIQ&sig2=0R5cHSsfGoE8nP_lkzRjLw&bvm=bv.49641647,d.dmg&cad=rja
Green for All (2010) Clean energy works Portland: a national model for energy-efficiency retrofits. Green for All, Oakland
Green Launching Pad (2011) Green launching pad. http://greenlaunchingpad.org
Laloudakis D (2011) Energize Phoenix: collecting and using data to improve the program Presentation at the What's Working in Residential Energy Efficiency Upgrade Programs Workshop, Washington. 20 May 2011
Lester P (2010) Clean Energy Works Portland: a model for retrofit projects. http://energy.gov/articles/clean-energy-works-portland-model-retrofit-projects
National Association of State Energy Officials (2011) U.S. state energy program briefing book. National Association of State Energy Officials, Alexandria
U.S. Department of Energy. Better Buildings Neighborhood Program (2013) Energizing efficiency upgrades in a Phoenix, Arizona, sustainability corridor. http://www1.eere.energy.gov/buildings/betterbuildings/neighborhoods/phoenix_profile.html
U.S. Environmental Protection Agency (2010) Clean Energy Works Portland: an energy efficiency retrofit program. EPA Local Climate and Energy Program Webcast. http://www.epa.gov/statelocalclimate/documents/pdf/transcript_smith_01-26-10.pdf
University of New Hampshire (2011) Green Launching Pad announces call for proposals for green manufacturing. http://www.unh.edu/news/cj_nr/2011/nov/lw14green.cfm
Venkatachalam AR (2011) Panel: Green Launching Pad. http://www.ebcne.org/fileadmin/pres/10-28-11_Master_-_Panel_Slides_-_NH_Chapter_Energy_and_Climate_Change_Program.pdf

Chapter 10
Common Themes and Conclusions

Abstract This concluding chapter explores important themes that emerged from the case studies and provides insights about the context in which EBED is implemented around the world. Some themes are more obvious, such as that EBED efforts tend to require multidimensional and comprehensive approaches, but others are more nuanced, such as that timing of EBED projects is crucial and often difficult. We also observe that EBED efforts may be met by unintended consequences when the benefits and burdens from energy development are distributed in unexpected ways and that community participation is important, especially for place-based approaches.

This chapter concludes that as the demands and challenges for low-carbon energy and for economic development continue into the 21st century, opportunities to successfully apply EBED will flourish. The framework established in this book, augmented with case study descriptions, is intended to help advance the knowledge base for EBED so that a community of research, practice, and policymaking grows along with the practice.

The case studies reviewed in the proceeding chapters demonstrate typical EBED dimensions, such as the pursuit of dual energy and economic goals; the use of low-emissions, efficient, and advanced energy technologies to achieve these goals; and the involvement of a diverse range of stakeholders. Additional commonalities across these cases warrant further examination. The common themes that we highlight in the discussion below lend additional insights about the context in which EBED is implemented and complement the EBED dimensions introduced in Chaps. 2 through 5, such as the process, framework, and policies that support EBED.

10.1 EBED Efforts Often Require a Multidimensional and Comprehensive Approach

Every case study discussed in Chaps. 7 through 9 embodies a strategic approach to EBED that incorporates multiple points of intervention. For example, some projects focus on product development and commercialization, such as for a solar lantern, but also seek to cultivate entrepreneurship to disseminate the lanterns and, as a result, reduce energy poverty in rural regions. This kind of program, accordingly, requires multiple skill sets in technology development, the use of household and cultural knowledge to help tailor the technology to specific communities' needs, entrepreneurship recruitment and training, financial models for dissemination, supportive policies, and the coordination of electricity provision efforts across a variety of actors. A large-scale, country-level strategy such as South Africa's SARi or ARRA in the United States elevates the degree to which the multidimensional aspects are required in EBED and the comprehensiveness of the EBED initiatives.

In addition to a multidimensional approach, EBED efforts tend to also be comprehensive because they entail a combination of social, environmental, and economic elements, as the case studies highlight. Related to our assertion early in this book that energy has become a driver of economic development rather than just an enabler, in EBED it is rare that energy is the intended and exclusive end goal of a project. Rather, energy is often a means to a higher-level end, such as the health of individuals, the health of an economy, environmental and ecosystem preservation, or economic opportunity. Therefore, energy advancement accomplishes little in EBED without also addressing issues of resource management, community empowerment and mobilization, and gender disparities, for example. In short, energy development should be pursued in conjunction with other development strategies, not in isolation. A more comprehensive approach is important to EBED success.

10.2 There is no Single Prescription

All EBED strategies are different based on context and scale of the initiative. Like any development effort, each EBED strategy needs to be tailored to its targeted community or market. In fact, as stressed in Chap. 3 one of the more important steps of the EBED process is assessing a locale's assets and how assets can be aligned for healthy growth in evolving and new markets. Thus, EBED strategies designed to take advantage of close-and moderate-alignment conditions, or to find ways to overcome limited-alignment conditions, are likely to be better positioned to successfully advance EBED goals. Because no two locations share the same conditions, we expect to find a plethora of different kinds of EBED practices as community and national EBED policy designers and implementers assess their locales' assets and their alignment with market opportunities.

There is also no one approach, or "silver bullet," that will solve energy and economic development challenges. Instead, our discussion on EBED is intended to offer researchers, practitioners, and policymakers additional tools to address some of the complex challenges we face in energy and economic development.

10.3 Timing is Crucial and Difficult

EBED efforts can entail several timing-related issues. We highlight three in particular. First, EBED can effectively target both short-term and long-term needs. We notice, however, that the time it takes to realize the return on an EBED investment often mirrors the level of transformation that the project aims to address, as first presented in Chap. 3. Initiatives that aim to create entirely new markets or products, and thus include higher levels of transformation, tend to take longer to attain desired impacts, whereas initiatives that link existing efforts will likely be able to generate impacts more quickly. It is important for evaluators and programs designers to understand the kind of EBED strategy they are undertaking and communicate the kinds of outcomes their program intends to achieve and the expected timeline. As the cases show, sometimes the short- and long-term goals are blended into one EBED initiative. Here, it is simply important for EBED program and policy designers to structure and communicate short- and long-term goals and objectives and how the program is meeting them.

For example, weatherization programs embedded in the Recovery Act were designed to provide immediate jobs and workforce training to address the rising unemployment during the recession. Existing and qualified construction workers, as well as others, were able to get training in energy efficiency construction techniques and get back to work relatively quickly. As these programs reach scale over time, the benefits from the program will shift to become greater for household and business owners that accrue savings from their reduced energy costs. Thus, this program had job creation impacts in the short term and energy savings impacts for households and businesses that grow over the medium and long term. This example illustrates that EBED strategies do not need to focus exclusively on short-term payoffs and can incorporate longer-term goals and objectives.

A second point on timing is that energy-related investments tend to experience a long-term return on investment. Many of the cases reviewed in the proceeding chapters entail significant investments in innovative technologies, services, or business models. The products and services that receive these investments may not, however, be commercially viable for years, if not decades later. Thus the financial, social, and environmental gains from energy-related investments are marked by longer time horizons. These investments therefore require "patient capital" or investment—investors understand that they must make short-term sacrifices in money expended to achieve significant long-term gains like greater financial returns, improved energy security, or reduced GHGs. For example, investments in new electric vehicle technologies may only create a handful of jobs in the short term, but these investments have the potential to seed new industries that employ thousands over the coming decades. These investments also may result in several longer-term social returns, such as the reduction of GHG emissions and enhanced energy security for nation-states.

Similarly, energy infrastructure investments are decades long, if not longer. A power plant, for example, may last between 30 and 60 years. Therefore, it is necessary to not only expect a longer time horizon for a return on energy investments, but decision-makers must also engage in long-term planning when considering energy infrastructure development. The implications of long-term investments like these are important because the infrastructure built today may outlive best practices in some future time period. It takes a long time to turn over infrastructure in particular, which makes it difficult to adopt new technologies that can improve this specific infrastructure. This difficulty in adopting new technologies to old infrastructure makes it easier to reinforce business-as-usual conditions and technological lock-in.

A final timing dimension involves the speed and extent to which countries are investing in new, innovative energy markets to establish themselves as a "first mover" in a niche energy area. This "great race" involves a number of countries scrambling to capture a dominant market position in the low-emissions, efficient, and advanced energy industries of the future, such as solar PV panels, electric vehicles, and algae biofuels. Thus, the race is less against the clock but more against each other. Market-based EBED strategies of this nature may not only involve significant risk and investment, but also a possibility of massive pay-off for long-term benefits such as increased GDP, job creation, and poverty alleviation.

10.4 Strategic Investment may be Necessary

Many cases reviewed in Chaps. 7 through 9 are instances where a country or a donor made strategic investments in a particular industry, business, or technology. Although many scholars advocate for neutrality in policy support for technologies or industries, others have documented that the successful transformation of emerging economies—including Japan, South Korea, and Taiwan—rests on a country making strategic targeted investments and crafting associated policies to support industry-specific growth. The majority of cases presented in this book mirror the latter approach of targeted, strategic investment in specific areas. Although we are not advocating for this approach over neutrality, it is important to recognize that strategic investment may be necessary when a specific industry, technology, or business is well positioned to successfully achieve EBED goals and objectives.

10.5 Project Self-Sufficiency can be Challenging

Many of the projects reviewed in the proceeding chapters were started with a large injection of government or donor funds. Notable exceptions are the social enterprise efforts such as Kamworks and Nuru Energy, which attracted supplemental donor or government funding, or support through procurement to expand their growth models. Regardless of the level of government or donor contributions, the challenge for these efforts is to become self-sustaining over time. Typically, programs achieve fiscal sustainability by leveraging other funds, particularly private industry support, or by effectively reaching markets and consumers willing to pay for the product or service at price points that sustain the operations and, if possible, growth of the endeavor. Aside from EBED efforts designed to be one-time injections of funds, other financial plans for self-sufficiency depend on the size, scale, and level of transformation that the initiative aims to accomplish. The range of business models to help launch and sustain EBED efforts abounds. Appropriate models for self-sufficiency depend on the financing, policy, government, and market environment that the EBED effort operates within. Similar to many new efforts trying to achieve scale and growth, keeping the effort focused on financially stable operations is critical for the EBED effort to maintain viability.

10.6 Public–Private Partnerships Play an Important Role

Given that the role of the private sector is often critical for EBED project sustainability, EBED often emerges as a type of public–private endeavor. This is especially true in market-oriented approaches, where the initiative targets business or industry development, yet these industries are nascent in their development and tend to need governmental support for investment, market protection, or some other form of policy to help them overcome market barriers. In many of the cases reviewed in the preceding chapters that involved public–private partnerships, the government provided financial support in the beginning phases of the project, and EBED project leaders used this support to leverage additional investment and participation from the private sector to scale the EBED effort.

Public–private partnerships are used increasingly to help business and industry overcome market failures such as high risks associated with medium- to long-term return on investment, where the private sector might not have enough patient capital to invest in the start-up and growth phases of product development. Other examples of market failure include a lack of information about markets or a business's competitors, which makes it difficult for businesses to make informed decisions about how to participate in competitive markets strategically. Another example of a market barrier is a business facing difficulty physically getting their product to markets because of poor transportation infrastructure. In these instances, the government can make a strong partner for business because it can leverage related infrastructure such as export routes.

10.7 Attention to Economic Benefit and Burden is Important

Most EBED efforts will affect specific subpopulations more than others. In fact, some EBED efforts, especially those that are household- and place-based approaches, are specifically designed for their intended recipients, such as low-income homeowners, the rural poor, women, children, the unemployed, or urban youth. It is not only important to at least consider who will benefit from an EBED effort, but also to consider whether an EBED effort burdens any subpopulations or individuals more than others. These considerations of justice, equity, and burden are often less prioritized than more streamlined components of the EBED process, but they are equally important to program outcomes.

10.8 EBED Efforts may be Met by Unintended Consequences

New initiatives, especially ones that are ground breaking or represent a significant shift from business-as-usual conditions, can produce unintended consequences. Such consequences may include disproportionate benefits and burdens to certain subpopulations, businesses, or industries or to certain aspects of society such the economy or the environment. Other unintended consequences may include the creation of perverse incentives, such as when a business sells inadequate or rip-off solar home system panels to a local market because of household-targeted solar panel incentives. In another example, energy development can occur at the cost of human or social development. This type of unintended consequence is evident in some of the case studies, such as the Lao PDR and Pennsylvania hydrologic fracturing cases. Both instances feature the inherent tensions between energy development and the environment, economic development, and social development. Both of these cases involve an attempt to develop a domestic low-carbon energy resource, with a main goal to increase domestic income and industry development. By exploiting the water and natural gas resources, both of these initiatives affect local ecosystems and potentially jeopardize the health and safety of individuals. The kinds of consequences exemplified in these two cases underscore the importance of a comprehensive approach to EBED, where these factors must be appropriately balanced, and energy should be considered a means to an end, not exclusively the end goal.

10.9 Political Will and a Consistent, Stable Policy Environment is Crucial

Two of the most common elements that emerge in all of the case studies are the presence of political will and a stable policy environment. Because policy plays such a prominent role in EBED efforts, it is crucial to EBED projects that the political and policy environment is stable and predictable. Inconsistent policies and funding streams can complicate, if not entirely thwart, long-term EBED project planning. On-again, off-again funding, which has historically been the case with the production tax credit in the United States, for example, makes it difficult to plan EBED projects and secure investor confidence, especially with projects that take a long time to complete or have significant upfront costs. A stable policy and political environment is particularly important when the EBED project is long term and requires years, if not decades, to evolve. Examples from the case studies for which political will over a long time horizon was noticeable important to EBED progress include Morocco's, Denmark's, and South Africa's renewable energy industry developments.

Because EBED efforts require more comprehensive approaches, EBED projects also often require large investments and several years, if not longer, to produce substantial project outcomes. These are investments that could be placed elsewhere to maintain a healthy citizenry and economy, such as in health system upgrades, nonenergy infrastructural developments, or social welfare. Although we are not asserting that EBED is more or less important than these alternative investments, we do note that an uneven political or policy environment that regularly bounces commitments from one sector to the other does not send a clear message to businesses, citizens, and other investors that are willing and interested in investing in energy developments.

10.10 Community Participation is Important, Especially for Place-Based Approaches

As emphasized in Chap. 3, stakeholder engagement is fundamental to the EBED process, especially at the stage in which EBED goals are defined and refined. When considering place-based and household-based EBED approaches in particular, one of the most important stakeholder groups to engage in a project is community members. Tailoring local programs or technologies to local needs and conditions is essential because individuals within a community will not embrace the effort if it threatens cultural or social norms, is inconvenient, or is incompatible with economic realities. The introduction of new, cleaner cookstoves is an example of a technology that has the potential to threaten cultural or social norms because a stove that uses nonbiomass fuel can directly change household dynamics. If it becomes no longer necessary for a woman or a child to gather firewood for cooking purposes, this time may be used to engage in other activities. For some, depending on the new activity, this type of change can threaten cultural norms such as girls spending more time in school or women performing tasks held by men. When introducing new products or services to a community, it is important to be aware of implications such as these. It is also important to assess whether and how community members can pay for the energy innovations, including not just payment for the product or service in aggregate terms, but also payment structures that are compatible with typical income patterns for the community.

Community participation is also important when the EBED effort involves new technology or other product uptake within a community. Technologies and other products need to be maintained over time, because they periodically break or suffer from vandalism. If those who own the products or selected community members are trained to maintain them, it is much more likely that these technologies or products will be used for a greater period of time. Energy efficiency and home weatherization efforts are relevant here as well. If a homeowner is trained to repair holes in siding, caulk windowsill leaks, or change the filter on a heating, ventilation, and air conditioning system, for example, she can continue to avoid energy losses over time.

10.11 EBED in the 21st Century

The demands for energy will only grow in the coming decades as countries continue to emerge from a developing status into middle-income economies, developed economies continue to require substantial energy inputs, and energy access is expanded in rural and poor regions. These trends, combined with the increasingly harmful effects of climate change and other forms of pollution, will continue to drive demand for low-emissions, efficient, and advanced energy. The energy development initiatives based on low-emissions and efficient energy are poised to provide many more sources of clean energy worldwide. These initiatives are also becoming important drivers for economic development. The billions of dollars invested in renewable energy and energy efficient developments each year by the public and private sectors are evidence that these needs exist and are likely here to stay for the foreseeable future.

This book offers a common framework and vocabulary for researchers, policymakers, and practitioners to take advantage of the confluence of this type of energy development, and the constant quest by governments to create jobs and transform economies. Applying a framework to these initiatives will help guide researchers, policymakers, and practitioners in designing and implementing more strategic and effective initiatives to meet the inherent challenges in energy and economic development. Understanding the suite of policies that can support EBED efforts may help enable or accelerate EBED efforts and, over time, lead to the creation of a policy environment more conducive to a range of EBED efforts.

10.11 EBED in the 21st Century

Expanding the types of methods and metrics used in EBED evaluations will help facilitate a stronger community of practice because it will inform others about lessons learned from a more holistic perspective. Finally, the themes documented in the case studies offer insights into how regions around the world are taking advantage of the opportunities inherent in this emerging and growing domain of EBED.